Skills Worksheet

Directed Reading A

Section: What Is Matter?

MATTER

Write the letter of the correct answer in the space provided.

_______ **1.** What do humans, hot soup, and a neon sign have in common?
- **a.** They are brightly colored.
- **b.** They are found in space.
- **c.** They are made of matter.
- **d.** They have the same volume.

_______ **2.** What has mass and takes up space?
- **a.** volume
- **b.** matter
- **c.** weight
- **d.** space

MATTER AND VOLUME

_______ **3.** What does the word *volume* refer to?
- **a.** an amount of matter
- **b.** an effect of gravity
- **c.** an amount of space
- **d.** an effect of mass

_______ **4.** Why can't another CD fit in a rack once the rack is completely filled?
- **a.** because all the space is taken up
- **b.** because the CD has mass
- **c.** because space has three dimensions
- **d.** because the rack has no volume

Liquid Volume

_______ **5.** Which of the following units would be best for expressing the volume of water in a lake?
- **a.** grams (g)
- **b.** liters (L)
- **c.** meters (m)
- **d.** milliliters (mL)

| Directed Reading A *continued*

______ **6.** Which of the following units would be best for expressing the volume
of soda in a can?
a. centimeters (cm)
b. grams (g)
c. liters (L)
d. milliliters (mL)

Measuring the Volume of Liquids

Use the terms from the following list to complete the sentences below.

  cubic   meniscus   volume

7. To measure volume with a graduated cylinder, look at the bottom of

the _________________________.

8. The volume of solid objects is usually expressed

in _________________________ units.

9. To find the _________________________ of a regular solid,
multiply its length, width, and height.

Volume of an Irregularly Shaped Solid Object

Use the terms from the following list to complete the sentences below.

  irregular solid   cubic centimeters   milliliter (mL)

10. To find the volume of a(n) _________________________,
measure the amount of water that the object displaces.

11. One cubic centimeter (1 cm^3) is equal to

one _________________________.

12. To express the volume of an irregular solid, you must change

milliliters to _________________________.

MATTER AND MASS

Write the letter of the correct answer in the space provided.

______ **13.** What is the measure of the amount of matter in an object?
a. matter
b. mass
c. volume
d. weight

Directed Reading A *continued*

_______ **14.** Which of the following is true about the mass of an object?
    a. Mass depends on the object's location.
    b. Mass is a measure of gravity.
    c. Mass does not depend on the object's location.
    d. Mass depends in part on weight.

_______ **15.** How could you change the mass of an object?
    a. Move it to the moon.
    b. Take some of its matter away.
    c. Make Earth spin faster.
    d. Change the object's weight.

The Difference Between Mass and Weight

_______ **16.** Which of the following is a measure of gravitational force?
    a. inertia
    b. mass
    c. volume
    d. weight

_______ **17.** What is the force that keeps objects from floating into space called?
    a. mass
    b. inertia
    c. gravitational force
    d. volume

_______ **18.** Which of the following is true about the weight of an object?
    a. Weight is measured with a balance.
    b. Weight is the same on the moon as on Earth.
    c. Weight is the same as mass.
    d. Weight depends on location in the universe.

Measuring Mass and Weight

_______ **19.** What is the weight on Earth of an object with a mass of 100 g?
    a. 1 newton
    b. 1 cm^2
    c. 1 mL
    d. 1 kilogram

Directed Reading A *continued*

Use the terms from the following list to complete the sentences below.

    newton   kilogram   mass

20. If a brick and a sponge have the same volume, the brick has

more _____________________.

21. The SI unit for mass is the _____________________.

22. The unit for weight is the SI unit for force called

the _____________________.

Skills Worksheet

Directed Reading A

Section: Physical Properties

Write the letter of the correct answer in the space provided.

______ **1.** Which of the following kinds of questions would be most useful for identifying objects?
   a. questions about their properties
   b. questions about their age
   c. questions about their weight
   d. questions about their volume

IDENTIFYING PHYSICAL PROPERTIES

______ **2.** What is a characteristic of an object that can be observed without changing the object's identity?
   a. a chemical property
   b. a flexible property
   c. a physical property
   d. a measurable property

Match the correct example with the correct property. Write the letter in the space provided.

______ **3.** aluminum flattened into thin sheets of foil

______ **4.** an ice cube made of solid water

______ **5.** copper pulled into thin wires

    a. ductility
    b. state
    c. malleability

Match the correct example with the correct property. Write the letter in the space provided.

______ **6.** flavored drink mix dissolving in water

______ **7.** a rose smelling sweet

______ **8.** a foam cup protecting your hand from a hot drink

    a. thermal conductivity
    b. solubility
    c. odor

Directed Reading A *continued*

Density

Write the letter of the correct answer in the space provided.

_______ **9.** Which physical property is the amount of matter in a given space, or volume?
   a. density
   b. ductility
   c. state
   d. weight

_______ **10.** If you know an object's mass (m) and volume (V), what equation would you use to find its density?

   a. $D = m \times V$ **c.** $D = \dfrac{V}{m}$

   b. $D = \dfrac{m}{V}$ **d.** $D = V + m$

_______ **11.** Which of the following units are most often used for the density of a solid?
   a. g/mL
   b. m^3/kg
   c. N/cm^3
   d. g/cm^3

Using Density to Identify Substances

_______ **12.** Which of the following is true of the density of each substance in the table titled Densities of Common Substances?
   a. The density makes the substance heavy.
   b. The density differs from the densities of other substances.
   c. The density is different at different times.
   d. The density is greater than the density of water.

_______ **13.** What is the density of lead shown in the table titled Densities of Common Substances?
   a. $1.00 \ g/cm^3$
   b. $0.0001663 \ g/cm^3$
   c. $13.55 \ g/cm^3$
   d. $11.35 \ g/cm^3$

_______ **14.** Which substance in the table titled Densities of Common Substances is a liquid that has a density greater than that of water?
   a. mercury
   b. ice
   c. helium
   d. lead

Directed Reading A *continued*

Density of Solids

_______ **15.** How does the volume of a kilogram of lead compare with the volume
of a kilogram of feathers?
   a. The lead has a larger volume.
   b. The lead has a smaller volume.
   c. The feathers have a smaller volume.
   d. The volumes are the same.

Density, Floating, and Sinking

_______ **16.** What happens to a solid object in water if its density is greater than
water?
   a. The object floats on top.
   b. The object dissolves.
   c. The object floats in the middle.
   d. The object sinks to the bottom.

Liquid Layers

_______ **17.** What causes different liquids to form layers when they are poured into
a container?
   a. the amounts of each liquid
   b. differences in density
   c. differences in color
   d. the temperatures of the liquids

_______ **18.** Where is the least dense liquid found when liquids form layers?
   a. in the lightest-colored layer
   b. in the middle layer
   c. floating on the top
   d. settled to the bottom

PHYSICAL CHANGES: NO NEW SUBSTANCES

Use the terms from the following list to complete the sentences below.

identity   physical change   state

19. Any change in matter that changes only its physical form is called

a(n) ____________________________.

20. All changes that cause a change of ____________________ are considered
physical changes.

21. When silver is molded into a pendant, its ____________________ is
the same.

Directed Reading A *continued*

Examples of Physical Changes

Write the letter of the correct answer in the space provided.

_______ **22.** Which of the following actions does NOT cause a purely physical change?
  a. sanding a piece of wood
  b. burning a piece of wood
  c. dissolving sugar in water
  d. freezing water

_______ **23.** Why is making ice from water a physical change?
  a. The ice has some new properties.
  b. The ice floats on water.
  c. The water changes its state.
  d. The water changes its identity.

Reversibility of Physical Changes

_______ **24.** Why are physical changes often easy to undo?
  a. The identity of the substance changes.
  b. The substance disappears.
  c. The identity of the substance does not change.
  d. Two new substances are made.

Matter and Physical Changes

_______ **25.** Why is making a star from a piece of paper considered a physical change?
  a. The paper's state has changed.
  b. The paper's identity is the same.
  c. The paper's color is the same.
  d. The paper has aged.

Skills Worksheet

Directed Reading A

Section: Chemical Properties
IDENTIFYING CHEMICAL PROPERTIES

Use the terms from the following list to complete the sentences below.

flammability

reactivity

nonflammability

chemical property

1. A property of matter that describes its ability to change into entirely

   new substances is called a(n) _______________________.

2. The ability of substances to change and form one or more new

   substances is a chemical property called _______________________.

3. The ability of a substance to burn is a chemical property

   known as _______________________.

4. Something that cannot burn has the property of _______________________.

Comparing Physical and Chemical Properties

Write the letter of the correct answer in the space provided.

_______ **5.** Which of the following sets of words describes only physical properties of a material?
   a. liquid, dense, flammable
   b. solid, ductile, yellow
   c. flammable, malleable, liquid
   d. powdery, reactive, insoluble

_______ **6.** What chemical property causes rust to form on a nail?
   a. conductivity
   b. nonflammability
   c. reactivity with oxygen
   d. flammability

_______ **7.** What do physical changes NOT change?
   a. the identity of the matter
   b. the amount of matter
   c. the state of matter
   d. the volume of the sample

_______ **8.** What makes chemical properties hard to observe?
   a. They cause changes of state.
   b. You can't see them until they produce new materials.
   c. Wearing protective glasses is required.
   d. They happen too quickly.

Directed Reading A *continued*

Characteristic Properties

_______ **9.** Which of these statements is true about characteristic properties of substances?
    a. They depend on sample size.
    b. They can only be physical properties.
    c. They can only be chemical properties.
    d. They can be physical properties or chemical properties.

CHEMICAL CHANGES AND NEW SUBSTANCES

_______ **10.** Which of these phrases describes a chemical change?
    a. pouring milk into a glass
    b. melting an ice cube
    c. digesting food in your body
    d. bending an iron nail

Use the terms from the following list to complete the sentences below.

    change   property

11. A chemical _____________________ describes which changes can happen to a substance.

12. A chemical _____________________ is a process by which substances actually change into new substances.

What Happens During a Chemical Change?

Write the letter of the correct answer in the space provided.

_______ **13.** Which of the following is an example of a chemical change?
    a. sugar dissolving
    b. a cake baking
    c. chocolate melting
    d. water freezing

_______ **14.** Which of the following describes what happens to the substances involved in a chemical change?
    a. The substances keep their identities.
    b. The substances change in form.
    c. The substances change into new substances with different properties.
    d. The substances combine and mix.

Directed Reading A *continued*

Signs of Chemical Changes

______ **15.** Which of the following usually happens during a chemical change?
     a. Heat is released or absorbed.
     b. The state of the matter changes.
     c. The identity of the matter stays the same.
     d. No heat is released or absorbed.

Matter and Chemical Changes

______ **16.** Why are chemical changes difficult to reverse?
     a. because they involve physical changes
     b. because they change the matter's form
     c. because they change the identity of the matter
     d. because their products are hard to find

PHYSICAL VERSUS CHEMICAL CHANGES

______ **17.** What is the type and the arrangement of the matter in an object called?
     a. the physical properties of the object
     b. the reactivity of the object
     c. the flammability of the object
     d. the composition of the object

______ **18.** How does a physical change differ from a chemical change?
     a. The change is not reversible.
     b. The composition of the matter is unchanged.
     c. New properties of the matter are created.
     d. New materials are produced.

______ **19.** What do chemical changes in a substance alter?
     a. the state of the substance
     b. the composition of the substance
     c. the size of the substance
     d. nothing in the substance

Reversing Changes

______ **20.** Why are chemical changes difficult to reverse?
     a. because they involve changes in composition
     b. because they involve changes in form
     c. because they involve changes in state
     d. because the temperature increases

Directed Reading B

Section: What Is Matter?

MATTER

1. What characteristic do a human, hot soup, the metal wires in a toaster, and the glowing gases in a neon sign have in common?

____

2. What is matter?

____

MATTER AND VOLUME

______ **3.** Which of the following units would be best for expressing the amount of water in a lake?
a. grams (g)
b. liters (L)
c. meters (m)
d. milliliters (mL)

______ **4.** Which of the following units would be best for expressing the volume of soda in a can?
a. centimeters (cm)
b. grams (g)
c. liters (L)
d. milliliters (mL)

5. What is volume?

____

6. Things with _________________________ cannot share the same space at the same time.

7. To measure the volume of water in a graduated cylinder, you should look at the bottom of the curve at the surface of the water called

the _____________________.

8. The volume of solid objects is commonly expressed

in _____________________ units.

9. What three dimensions are needed to find the volume of a rectangular solid?

____

____

Directed Reading B *continued*

---

10. How could the volume of a 12-sided object be found using water and a graduated cylinder?

____

____

11. If the volume of water displaced by the 12-sided object is 8 mL, what is the volume of the 12-sided object in cubic units?

____

MATTER AND MASS

______ **12.** The measure of the amount of matter in an object is its
  a. volume.
  b. length.
  c. meniscus.
  d. mass.

______ **13** The measure of the gravitational force on an object is its
  a. mass.
  b. length.
  c. weight.
  d. volume.

______ **14.** The SI unit of mass is the
  a. newton.
  b. liter.
  c. kilogram.
  d. pound.

______ **15.** One newton is about equal to the weight of an object that has
  a. a mass of 100 g on the moon.
  b. a volume of 1 m^3 on Earth.
  c. a mass of 1 kg on Earth.
  d. a mass of 100 g on Earth.

16. What is the only way to change the mass of an object?

____

____

Directed Reading B *continued*

For each description, write whether it applies to mass or to weight.

_____________________ **17.** is always constant no matter where the object is located

_____________________ **18.** is measured using a spring scale

_____________________ **19.** is expressed in grams (g), kilograms (kg), or milligrams (mg)

_____________________ **20.** is expressed in newtons (N)

_____________________ **21.** is less on the moon than on Earth

Skills Worksheet

Directed Reading B

Section: Physical Properties
IDENTIFYING PHYSICAL PROPERTIES

_______ **1.** A characteristic of matter that can be observed or measured without changing the identity of the matter is a
a. matter property.
b. physical property.
c. chemical property.
d. volume property.

_______ **2.** Some examples of physical properties are
a. color, odor, and reactivity.
b. color, odor, and speed.
c. color, odor, and mass.
d. color, odor, and anger.

Match the correct example with the correct physical property. Write the letter in the space provided.

_______ **3.** Aluminum can be flattened into sheets of foil.

_______ **4.** Water is frozen into ice.

_______ **5.** Copper can be pulled into thin wires.

_______ **6.** Your hand grows warm from holding a cup of hot liquid.

_______ **7.** Flavored drink mix dissolves in water.

_______ **8.** An onion gives off a very distinctive smell.

_______ **9.** A golf ball has more mass than a table tennis ball.

a. state
b. solubility
c. thermal conductivity
d. malleability
e. odor
f. ductility
g. density

10. Density is the _____________________ that describes the relationship between mass and volume.

11. The amount of matter in a given space, or volume, is called

_______________________.

12. What is the equation for density?

____

Directed Reading B *continued*

13. What do *D*, *V*, and *m* stand for in the equation for density?

____

____

____

14. The units for density consist of a mass unit divided by a(n)

_______________________ unit.

15. What happens to the density of a given substance if you increase the amount of the substance that you have?

____

16. What are two reasons why density is a useful physical property for identifying substances?

____

____

____

17. Why would 1 kilogram of lead be less awkward to carry around than 1 kilogram of feathers?

____

____

____

18. What will happen to a solid object made from matter with a greater density than water when it is dropped into water?

____

____

19. How will knowing the density of a substance help you determine whether an object made from that material will float in water?

____

____

20. If you pour different liquids into a graduated cylinder, the liquids will form

layers based upon differences in their _______________________.

Directed Reading B *continued*

21. If you pour different liquids into a graduated cylinder, which layer of liquid will settle on the bottom?

____

22. If you pour different liquids into a graduated cylinder, where will the layer of liquid with the lowest density be found?

____

PHYSICAL CHANGES: NO NEW SUBSTANCES

23. A change that affects only the physical properties of a substance is

known as a(n) _________________.

24. What kind of changes are changes of state, such as melting and freezing?

____

Identify which of the following activities represent physical changes by writing *PC* in the space provided. Put an *X* beside activities that do not.

_______ **25.** sanding a piece of wood

_______ **26.** baking bread

_______ **27.** crushing an aluminum can

_______ **28.** melting an ice cube

_______ **29.** dissolving sugar in water

_______ **30.** molding a piece of silver

31. When a substance undergoes a physical change, its _________________ does not change.

32. What is changed when matter undergoes a physical change? Give an example to explain your answer.

____

____

____

____

Directed Reading B

Section: Chemical Properties
IDENTIFYING CHEMICAL PROPERTIES

______ **1.** A property of matter that describes its ability to change into new matter with different properties is known as a(n)
  a. chemical change.
  c. chemical property.
  b. physical change.
  d. physical property.

______ **2.** The chemical property that describes the ability of substances to change and form one or more new substances is called
  a. reactivity.
  c. density.
  b. flammability.
  d. solubility.

______ **3.** The ability of a substance to burn is a chemical property known as
  a. ductility.
  c. density.
  b. flammability.
  d. solubility.

______ **4.** An iron nail is reactive with
  a. rubbing alcohol.
  b. other iron nails.
  c. wood in a house.
  d. oxygen in the air.

______ **5.** Which of the following statements is true about characteristic properties of matter?
  a. Characteristic properties depend on the size of the sample.
  b. Characteristic properties may be either physical or chemical properties.
  c. Characteristic properties involve only chemical properties.
  d. Characteristic properties involve only the physical nature of the matter.

6. Describe how burning changes the nature of wood.

____

____

7. Observing the _______________________ properties of a substance involves changing the identity of the substance.

8. The properties that are most useful in identifying a substance are

called _______________________ properties.

Directed Reading B *continued*

CHEMICAL CHANGES AND NEW SUBSTANCES

_______ **9.** Chemical changes are the processes by which substances
      a. move from place to place.
      b. change into new substances.
      c. change their physical properties.
      d. become greater in mass.

_______ **10.** Which of the following would NOT be considered an example of a chemical change?
      a. the bubbling action of effervescent tablets
      b. the formation of green coating on copper statues
      c. the melting of an ice cream bar
      d. the burning of rocket fuel

11. How do you know that baking a cake involves chemical changes?

____

____

____

____

12. List some signs or clues that show that a change you are observing is a chemical change.

____

____

____

____

13. An increase in the surrounding temperature is felt when a chemical

change _______________________ heat.

14. A decrease in the surrounding temperature is felt when a chemical

change _______________________ heat.

15. Because _______________________ changes cause a change in the identity of the substances involved, they are hard to reverse.

❙ Directed Reading B *continued*

16. How could some chemical changes be reversed? Give an example.

____

____

____

____

PHYSICAL VERSUS CHEMICAL CHANGES

_______ **17.** What is the most important question to ask to determine whether a change is physical or chemical?
  a. Was there a color change?
  b. Did the composition change?
  c. Was there a change in size?
  d. Did the change involve a change in state?

_______ **18.** The composition of a substance does not change during
  a. physical changes.
  b. chemical changes.
  c. reactivity.
  d. reversibility.

_______ **19.** The chemical changes that happen when a firework explodes are
  a. physical changes.
  b. easily reversed.
  c. almost impossible to reverse.
  d. changes only in state.

Identify whether the following changes are physical changes or chemical changes. Label each change either *PC* for physical change or *CC* for chemical change.

_______ **20.** effervescent tablets bubbling in water

_______ **21.** grinding baking soda into a powder

_______ **22.** souring milk

_______ **23.** freezing water into ice cubes

_______ **24.** burning a wooden match

_______ **25.** mixing drink mix into water

_______ **26.** bending an iron nail

Skills Worksheet

Vocabulary and Section Summary A

What Is Matter?

VOCABULARY

In your own words, write a definition of the following terms in the space provided.

1. matter

2. volume

3. meniscus

4. mass

5. weight

SECTION SUMMARY

Read the following section summary.

- Two properties of matter are volume and mass.
- Volume is the amount of space taken up by an object.
- Mass is a measure of the amount of matter in an object.
- The SI unit of volume is the liter (L). The SI unit of mass is the kilogram (kg).
- Weight is a measure of the gravitational force on an object, usually in relation to Earth. Weight is expressed in newtons (N).

Skills Worksheet

Vocabulary and Section Summary A

Physical Properties

VOCABULARY

In your own words, write a definition of the following terms in the space provided.

1. physical property

____

____

____

2. density

____

____

____

3. physical change

____

____

____

SECTION SUMMARY

Read the following section summary.

- Physical properties of matter can be observed without changing the identity of the matter.
- Examples of physical properties are melting temperature, density, hardness, thermal conductivity, and electrical conductivity.
- Density is the amount of matter in a given space.
- Density can be used to identify substances because the density of a substance is constant at a given pressure and temperature.
- When a substance undergoes a physical change, its identity stays the same.
- Physical changes include dissolving, cutting, bending, freezing, and melting.

Skills Worksheet

Vocabulary and Section Summary A

Chemical Properties

VOCABULARY

In your own words, write a definition of the following terms in the space provided.

1. chemical property

____

____

____

2. chemical change

____

____

____

SECTION SUMMARY

Read the following section summary.

- Chemical properties describe the ability of a substance to change into a new substance.
- The chemical properties of a substance describe how the substance will behave under conditions that favor a chemical change.
- Reactivity and flammability are chemical properties.
- New substances form as a result of a chemical change.
- Chemical changes usually liberate or absorb heat.
- Chemical changes alter the composition of a substance.

Vocabulary and Section Summary B

What Is Matter?

VOCABULARY

After you finish reading the section, try this puzzle! Use the clues below to solve the crossword puzzle.

ACROSS

2. the curve at a liquid's surface by which one measures the volume of the liquid

3. a measure of the gravitational force exerted on an object; its value can change with the location of the object in the universe

4. a measure of the size of a body or region in three-dimensional space

DOWN

1. a measure of the amount of matter in an object

2. anything that has mass and takes up space

SECTION SUMMARY

Read the following section summary.

- Two properties of matter are volume and mass.
- Volume is the amount of space taken up by an object.
- Mass is a measure of the amount of matter in an object.
- The SI unit of volume is the liter (L). The SI unit of mass is the kilogram (kg).
- Weight is a measure of the gravitational force on an object, usually in relation to Earth. Weight is expressed in newtons (N).

Skills Worksheet

Vocabulary and Section Summary B

Physical Properties

VOCABULARY

After you finish reading the section, try this puzzle! In the space provided, write the term described. Then, find the words in the word search puzzle on the next page. Words are hidden vertically, horizontally, diagonally, and backward.

____________________ **1.** a characteristic of a substance that does not involve a chemical change, such as density, color, or hardness

____________________ **2.** the ratio of the mass of a substance to the volume of the substance

____________________ **3.** a change of matter from one form to another without a change in chemical properties

____________________ **4.** a measure of the size of a body or region in three-dimensional space

____________________ **5.** a measure of the amount of matter in an object

SECTION SUMMARY

Read the following section summary.

- Physical properties of matter can be observed without changing the identity of the matter.

- Examples of physical properties are melting temperature, density, hardness, thermal conductivity, and electrical conductivity.

- Density is the amount of matter in a given space.

- Density can be used to identify substances because the density of a substance is constant at a given pressure and temperature.

- When a substance undergoes a physical change, its identity stays the same.

- Physical changes include dissolving, cutting, bending, freezing, and melting.

Vocabulary and Section Summary B *continued*

W	I	M	M	Q	W	T	I	D	I	R	X	D	P	V	Y	Y	C	H	H
J	T	A	S	Y	T	V	F	R	B	Q	T	L	E	T	Y	R	Y	D	H
K	S	L	O	K	L	Q	P	Z	W	D	X	S	R	N	Z	G	O	X	B
S	B	C	N	R	C	U	I	W	C	F	G	E	A	D	S	V	H	N	A
F	H	C	S	G	H	C	F	P	U	C	P	T	M	R	X	I	A	U	A
R	R	D	G	G	P	V	S	M	K	O	R	O	Z	X	I	P	T	O	W
Z	H	M	Z	D	S	Y	Q	C	R	C	E	A	B	E	C	E	I	Y	N
B	S	Q	C	B	M	Y	J	P	A	R	K	E	I	Y	N	I	P	B	P
E	Z	N	O	K	G	V	L	Q	L	M	L	Y	M	Z	N	N	L	B	O
L	Q	U	O	L	D	A	E	B	G	Y	F	W	D	R	Z	R	L	Q	V
G	P	S	B	M	C	L	X	S	B	L	Z	Q	E	P	G	X	O	I	U
G	R	Q	X	I	A	K	Q	O	E	Q	F	N	Q	I	X	H	Y	W	H
P	Y	K	S	G	A	W	U	X	B	S	Z	K	S	K	B	V	A	Q	U
A	D	Y	Q	X	F	X	I	C	K	M	I	B	W	I	I	O	O	X	W
D	H	P	H	Y	S	I	C	A	L	C	H	A	N	G	E	L	D	M	K
P	X	B	H	F	T	S	X	B	S	Z	D	P	P	B	X	U	N	M	O
A	O	H	P	P	W	N	O	G	U	J	N	A	O	B	Q	M	F	P	M
E	P	U	X	A	Z	G	D	P	R	D	V	T	S	T	B	E	B	W	A
E	K	G	B	U	P	W	D	G	M	P	S	C	H	O	A	X	Y	J	L
M	B	X	P	A	S	H	K	U	Z	C	N	Y	E	U	Y	P	V	J	X

Skills Worksheet

Vocabulary and Section Summary B

Chemical Properties

VOCABULARY

After you finish reading the section, try this puzzle! Unscramble the letters at the end of each description to find the term that is described.

1. a property of matter that describes a substance's ability to participate in chemical reactions: ERYOCMRHLPEPCITA

___ ___ ___ ___ ___ ___ ___ ___   ___ ___ ___ ___ ___ ___ ___ ___

2. the ability of a substance to change into one or more new substances: TECRIYAVTI

___ ___ ___ ___ ___ ___ ___ ___ ___ ___

3. a change that occurs when one or more substances change into entirely new substances with different properties: CILACHHGECNAME

___ ___ ___ ___ ___ ___ ___ ___   ___ ___ ___ ___ ___ ___

4. constant properties that are most useful in identifying a substance: TICOTICSRIAHCEPRPTRAEESR

___ ___ ___ ___ ___ ___ ___ ___ ___ ___ ___ ___

___ ___ ___ ___ ___ ___ ___ ___ ___ ___ ___ ___

5. the ability of a substance to burn: YMALMFITIBAL

___ ___ ___ ___ ___ ___ ___ ___ ___ ___ ___ ___

6. the type of matter that makes up an object and its arrangement in the object: CPIMSONOITO

___ ___ ___ ___ ___ ___ ___ ___ ___ ___ ___

SECTION SUMMARY

Read the following section summary.

- Chemical properties describe the ability of a substance to change into a new substance.
- The chemical properties of a substance describe how the substance will behave under conditions that favor a chemical change.
- Reactivity and flammability are chemical properties.
- New substances form as a result of a chemical change.
- Chemical changes usually liberate or absorb heat.
- Chemical changes alter the composition of a substance.

Name _______________________________ Class _______________ Date _______________

Reinforcement

A Matter of Density

Complete this worksheet after you finish reading the section "Physical Properties."

Imagine that you work at a chemical plant. This morning, four different liquid chemicals accidentally spilled into the same tank. Luckily, none of the liquids reacted with one another! Also, you know the liquids do not dissolve in one another, so they must have settled in the tank in four separate layers. The sides of the tank are made of steel, so you can see only the surface of what's inside. But you need to remove the red chemical to use in a reaction later this afternoon.

How will you find and remove the red chemical? By finding the chemicals' different densities, of course!

The following liquids were spilled into the tank:

- a green liquid that has a volume of 48 L and a mass of 36 kg
- a blue liquid that has a volume of 144 L and a mass of 129.6 kg
- a red liquid that has a volume of 96 L and a mass of 115.2 kg
- a black liquid that has a volume of 120 L and a mass of 96 kg

1. Calculate the density of each liquid.

Green liquid: ____

Blue liquid: ____

Red liquid: ____

Black liquid: ____

2. Determine the order in which the liquids have settled in the tank.

First (bottom): ____

Second: ____

Third: ____

Fourth (top): ____

_______ **3.** What kind of property did you use to distinguish among these four chemicals?
  a. a chemical property
  b. a physical property
  c. a liquid property
  d. a natural property

Reinforcement *continued*

4. Use colored pencils to sketch and label the position of the liquid layers in the tank on the diagram shown below.

5. Now that you know where the red chemical is inside the tank, how would you remove it?

____

____

____

____

Critical Thinking

As a Matter of Fact!
FROM THE JOURNAL OF CAPTAIN JANE P. FLEET

LOG 2551

I have sent two of my best science officers to explore the planet Xerxes. Their mission is to collect samples of matter from the planet's surface.

LOG 2552

The science officers brought back a small cube of space matter from Xerxes. The cube is white, odorless, about the size of a grapefruit, and it glows in the dark. We will observe the cube for a few days.

LOG 2553

Last night, the space cube expanded to three times its normal size. It is now about the size of a packing box. Its mass did not change.

LOG 2554

Today the cube divided into four smaller cubes. Each new cube has more mass than the original cube.

LOG 2555

Lab officers applied electricity to one of the cubes. The cube burst into flames and exploded, covering the room and our science officers with a green paste. The paste reacted with the surfaces, and now the walls and the science officers are permanently green.

COMPREHENDING IDEAS

1. What changes in the properties of the space cube were recorded in Log entry 2553?

____

____

2. Explain how the changes recorded in Log entry 2553 affected the space cube's density.

____

____

____

____

Critical Thinking *continued*

ANALYZING INFORMATION

3. List the properties of the cube that changed between Log 2551 and Log 2555.

____

____

4. Describe whether each change was chemical or physical.

____

____

____

COMPARE AND CONTRAST

5. Which of the cube's changes are not commonly observed in matter found on Earth? Explain your answer.

____

____

____

____

6. Which of the cube's changes might be possible to observe in matter on Earth? Explain your answer.

____

____

____

____

____

SciLinks Activity

DESCRIBING MATTER

Go to www.scilinks.org. To find links related to Describing Matter, type in the keyword HY70391. Then, use the links to answer the following questions about describing matter.

Internet Resources

For a variety of links related to this chapter, go to www.scilinks.org

Topic: Describing Matter
SciLinks code: HY70391

1. What are materials scientists and what do they do?

____

____

____

____

2. List five broad categories of materials studied by materials scientists. Describe their properties and how their uses relate to their properties.

____

____

____

____

____

3. Describe five properties of water. How do its properties help explain why water is such a unique substance?

____

____

____

____

____

____

Name _________________________________ Class _______________ Date _____________

Section Review

What Is Matter?

USING VOCABULARY

1. Use *volume* and *meniscus* in the same sentence.

____

____

____

2. Write an original definition for *matter*, *mass*, and *weight*.

____

____

____

UNDERSTANDING CONCEPTS

3. Identifying What units can be used to express the volume of a solid?

____

____

____

____

____

CRITICAL THINKING

4. Evaluating Data A nugget of gold is placed in a graduated cylinder that contains 85 mL of water. The water level rises to 225 mL after the nugget is added to the cylinder. What is the volume of the gold nugget?

____

____

____

____

5. Identifying Relationships Do objects that have large masses always have large weights? Explain your answer.

____

____

____

____

____

____

MATH SKILLS

Interpreting Graphics

The table below lists measurements for three blocks of wood. Use the table to answer the next two questions.

Wood block	Area of the base (m²)	Height (m)
B1	36	4
B2	16	9
B3	A	12

6. Analyzing Shapes Find the volume of B1 and of B2. Show your work below.

7. Analyzing Shapes If the three blocks have equal volumes, what is A, the area of the base of B3? Show your work below.

Section Review *continued*

CHALLENGE

8. Applying Concepts Compare an elephant's weight on the moon with its weight on Earth. Explain your answer.

____

____

____

____

____

____

____

____

Section Review

Physical Properties

USING VOCABULARY

1. Use *physical property* and *physical change* in separate sentences.

UNDERSTANDING CONCEPTS

2. Comparing Explain why a golf ball is heavier than a table-tennis ball even though the balls are the same size.

3. Describing Explain what happens to a substance when it goes through a physical change.

Section Review *continued*

INTERPRETING GRAPHICS

Use the table below to answer the next two questions.

Substance	Density* (g/cm³)
Zinc (solid)	7.13
Silver (solid)	10.50
Lead (solid)	11.35

*at 20°C and 1.0 atm

4. Identifying Suppose that 273 g of one of the substances listed above displaces 26 mL of water. What is the substance?

____

____

____

____

5. Evaluating How many milliliters of water would be displaced by 408 g of lead?

____

____

____

____

CRITICAL THINKING

6. Applying Concepts How can you determine that a coin is not pure silver if you know the mass and volume of the coin?

____

____

____

____

____

____

Section Review *continued*

7. Identifying Relationships What physical property do water, oil, mercury, and alcohol share?

____

____

____

____

____

MATH SKILLS

8. Using Equations What is the density of an object that has a mass of 350 g and a volume of 95 cm^3? Will the object float in water? Show your work below.

CHALLENGE

9. Analyzing Processes Write a step-by-step process for finding the density of an unknown liquid. List the laboratory equipment needed for each step.

____

____

____

____

____

____

____

____

Section Review

Chemical Properties

USING VOCABULARY

1. Write an original definition for *chemical property* and *chemical change*.

UNDERSTANDING CONCEPTS

2. Comparing Explain why rusting is a chemical change and not a physical change.

3. Describing Write two examples of chemical properties, and explain why they are chemical properties.

4. Applying Originally, the Statue of Liberty was copper colored. After being exposed to air, she turned green. What kind of change happened? Explain.

| Section Review *continued*

---

5. Summarizing What are two ways that heat can be involved in a chemical change?

____

____

____

____

____

CRITICAL THINKING

6. Making Comparisons Describe the difference between physical and chemical changes in terms of what happens to the matter involved in each kind of change.

____

____

____

____

____

____

7. Applying Concepts Is melting a physical change or a chemical change? Explain your answer.

____

____

____

____

____

____

____

Section Review *continued*

MATH SKILLS

8. Evaluating Data The temperature of an acid solution is 25°C. A strip of magnesium is added to the solution, and the temperature rises 2°C per min for the first 3 min. After an additional 5 min, the temperature has risen another 2°C. What is the final temperature of the solution? Show your work below.

CHALLENGE

9. Analyzing Processes As a candle burns, chemical changes take place. Describe two of these changes, and explain how you know that these changes are chemical changes and not physical changes.

____

____

____

____

____

____

____

____

____

____

____

____

Chapter Review

USING VOCABULARY

1. Academic Vocabulary In the sentence, "Chemical changes usually liberate or absorb heat," what does the word *liberate* mean?

____

____

2. Use *physical property*, *chemical property*, *physical change*, and *chemical change* in separate sentences.

____

____

____

____

For each pair of terms, explain how the meanings of the terms differ.

3. *mass* and *weight*

____

____

____

4. *volume* and *density*

____

____

____

UNDERSTANDING CONCEPTS
Multiple Choice

_______ **5.** Which of the following properties is a physical property?
   a. reactivity with oxygen
   b. malleability
   c. flammability
   d. reactivity with acid

Chapter Review *continued*

______ **6.** Volume can be expressed in any of the following units EXCEPT
     a. grams.
     b. liters.
     c. milliliters.
     d. cubic centimeters.

______ **7.** What is the SI unit for mass?
     a. gram
     b. liter
     c. milliliter
     d. kilogram

______ **8.** The best way to measure the volume of an irregularly shaped solid is
to use
     a. a ruler to measure the length of each side of the object.
     b. a balance.
     c. the water displacement method.
     d a spring scale.

______ **9.** Which of the following statements about density is true?
     a. Density is expressed in grams.
     b. Density is mass per unit volume.
     c. Density is expressed in milliliters.
     d. Density is a chemical property.

Short Answer

10. Listing List two characteristic properties of matter.

____

____

____

____

11. Comparing Explain how the process of measuring a liquid's volume differs
from the process of measuring a solid's volume.

____

____

____

____

Chapter Review *continued*

INTERPRETING GRAPHICS

Use the image below to answer the next two questions.

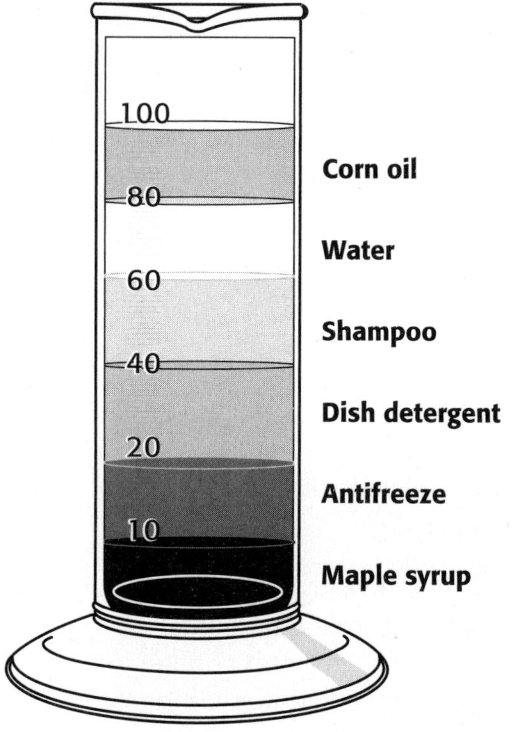

12. Analyzing How does the density of dish detergent compare with the density of water and maple syrup?

____

____

____

____

13. Applying How would the layers appear in a graduated cylinder that contained only shampoo, corn oil, and antifreeze?

____

____

____

____

____

____

INTERPRETING GRAPHICS

Use the illustration below to answer the next four questions.

14. Listing List three physical properties of this aluminum can.

____

____

____

15. Classifying When this can was crushed, did it undergo a physical change or a chemical change?

____

____

____

16. Analyzing How does the density of the metal in the crushed can compare with the metal's density before the can was crushed?

____

____

____

17. Concluding Can you determine the can's chemical properties by looking at the picture? Explain your answer.

____

____

____

____

| Chapter Review *continued*

WRITING SKILLS

18. Technical Writing Write a set of instructions that describe how to find the density of an object. Write the instructions so that they work for a regularly shaped object and for an irregularly shaped object. List the materials needed and the sequence of steps to follow.

CRITICAL THINKING

19. Concept Mapping Use the following terms to create a concept map: *matter, mass, volume, milliliters,* and *cubic centimeters.*

20. Making Inferences You mix two substances in a beaker and expect them to undergo a chemical change. The temperature of the mixture does not change. Has a chemical change occurred? Explain your answer.

____

____

____

____

| Chapter Review *continued*

INTERPRETING GRAPHICS

Use the graph below to answer the next two questions.

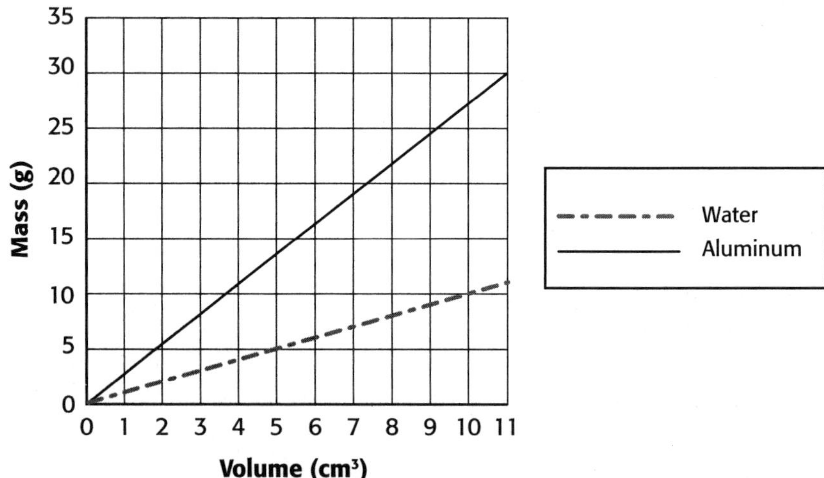

21. Identifying Relationships What do the slopes of each line represent?

____

____

____

22. Applying Concepts Which substance has the greater density?

____

____

____

MATH SKILLS

23. Using Equations What is the volume of a book that has a width of 10 cm, a length that is 2 times the width, and a height that is half the width? Express your answer in cubic units. Show your work below.

❙ Chapter Review *continued*

---

24. Using Equations A jar contains 30 mL of glycerin (whose mass is 37.8 g) and 60 mL of corn syrup (whose mass is 82.8 g). Which liquid is on top? Explain your answer. Show your work below.

CHALLENGE

25. Analyzing Processes You are making breakfast for your friend Filbert. When you take the scrambled eggs to the table, he asks, "Would you please poach these eggs instead?" Use the ideas of chemical change and physical change to give Filbert a scientific reason that you are unable to poach the eggs.

____

____

____

____

____

____

____

____

____

____

____

____

____

____

____

Assessment

Chapter Pretest

Teacher Notes and Answer Key

The Pretest questions are designed to help you determine the prior knowledge of your students. Some questions test whether students have mastered the background knowledge they need to understand the content you are about to teach. Other questions test your students' prior knowledge of the content you are about to teach. Use the Pretest with the Test Doctors and diagnostic teaching tips in these notes pages to help you tailor your instruction to your students' specific needs.

QUESTION NUMBER	CORRECT ANSWER	STANDARD
1	B	8.9.b
2	A	8.9.c
3	A	8.9.d
4	C	8.9.f
5	D	8.9.g
6	D	8.5.a
7	C	8.8.a
8	C	8.8.b
9	B	8.8.d

TEST DOCTOR

The following Pretest questions have been diagnosed by the Test Doctor. Find out what might be causing your students' "ailing" answers. Each Test Doctor is followed by a diagnostic teaching tip to help you address students' learning needs.

Question 1 *asks students to determine if a measurement will be inaccurate.*

A Incorrect. This is an accurate measurement.

B Correct. This is not an accurate measurement. Error will occur if a tool with such gross units is used. A graduated cylinder should be used instead.

C Incorrect. This is an accurate measurement.

D Incorrect. This is an accurate measurement.

Diagnostic Teaching Tip: Students who have difficulty with this question should review accuracy in measuring, including choosing the right tool with which to measure. Ask the class why they think a measuring cup is not accurate enough to give a reliable measurement in science. Discuss why it is OK to use when cooking (measurements do not have to be as precise). Remind students that besides using the correct tool, one must also read the measurement in the correct way.

Question 2 *asks students to determine the controlled parameters of an experiment.*

A Correct. Varying who enters the room would test the response of the dog to that particular person.

B Incorrect. The number of people in the room should be a controlled parameter.

Chapter Pretest *continued*

C Incorrect. The noise level in the room should be a controlled parameter.
D Incorrect. The time of day you run your tests should be a controlled parameter.

Diagnostic Teaching Tip: Students who have difficulty with this question might benefit from reviewing each condition with a partner. Have partners discuss why each incorrect answer should be a controlled parameter. Then, have them plan how they would control each one.

Question 3 *asks students to determine the slope, given a set of data from a linear graph.*

A Correct. The slope of the line will be the change in the y value over the change in the x value, or 8/5.
B Incorrect. This is the initial y value over the last y value.
C Incorrect. This is the last x value over the initial x value.
D Incorrect. This is the x value over the y value before it is put in the simplest terms.

Diagnostic Teaching Tip: Students who have difficulty with this question should review how to find the slope from a linear graph. Plot the given data points for the students, and draw the line containing the points. Remind students that the slope of a line is found by taking the change in the y value over the change in the x value. Any two points can be used to find these values, but the number must written in simplest terms. Show students that the slope is the same by giving two examples: the change from (0,0) to (5,8) and the change from (0,0) to (25,40). In this case, the change in values is 8/5, which is the slope of the line.

Question 4 *asks students to find the volume of a substance, given its density and mass.*

A Incorrect. In this case, the student has used the wrong equation for density, or has used $d = mv$ instead of using $d = m/v$.
B Incorrect. Here, the student has used the same incorrect equation as used in Choice A and also has misplaced the decimal point.
C Correct. The equation for density is density equals mass over volume. This equation was solved correctly for volume.
D Incorrect. In this case, the student has used the wrong equation for density, or has used $d = v/m$ instead of using $d = m/v$.

Diagnostic Teaching Tip: Students who have trouble answering this question correctly should work with a partner to solve the density equation for volume. Have partners solve for volume and then use the given data to find the volume of the bell. If student pairs are having difficulty, help them with the calculation, or partner them with a third student who excels in mathematics.

Question 5 *asks students to determine if a graph is linear or nonlinear with a direct or an inverse relationship.*

A Incorrect. Losing weight at a lesser rate later in the diet is not a linear relationship. Also, losing weight as time advances is not a direct relationship.

Chapter Pretest *continued*

B Incorrect. Losing weight at a lesser rate later in the diet is not a linear relationship.

C Incorrect. Losing weight as time advances is not a direct relationship.

D Correct. Losing weight at a lesser rate later in the diet is a nonlinear relationship. Also, losing weight as time advances is an inverse relationship.

Diagnostic Teaching Tip: Students who have difficulty answering this question correctly should be shown the four types of graphs described in this question. Ask student volunteers what each graph shows (linear or nonlinear graph; direct or inverse relationship). After students have reviewed the graphs, return to the question, and discuss why it is a nonlinear graph showing an inverse relationship.

Question 6 *asks students to identify a chemical change.*

A Incorrect. Molecules change in a chemical reaction.

B Incorrect. A molecule's movement varies with temperature, not age.

C Incorrect. A chemical change forms a new substance, not crystals of the same substance.

D Correct. Interaction by molecules in a chemical change forms a new product.

Diagnostic Teaching Tip: Students who have difficulty answering this question correctly might benefit from an introduction to the difference between a physical and chemical change. Tell students to think of two situations involving milk. In one case, a person accidentally puts a carton of milk in the freezer instead of the refrigerator. Tell students that the milk will freeze, and that this is only a change in the state of matter, which is a physical change. Then tell them that in the second case, the person accidentally put the milk in a pantry, where after a week it began to smell. Tell them that the smell means a chemical change, or one that changes the substance into a new one, has taken place. Be sure to emphasize Section 3, "Chemical Properties," in Chapter 3, "Properties of Matter." Students must know that reactant atoms and molecules interact to form products with different chemical properties in order to master standard 8.5.a.

Question 7 *asks students to know that a substance's density remains the same regardless of how much of it is there.*

A Incorrect. Density of a substance does not change regardless of mass.

B Incorrect. Density of a substance does not change regardless of mass.

C Correct. Density of a substance is its mass divided by its volume and does not change.

D Incorrect. Density of a substance does not change regardless of mass. Its volume will be in proportion to its mass.

Diagnostic Teaching Tip: Some students having difficulty with this question should be introduced to two differently shaped forms of the same substance, such as a copper flowerpot and a copper frying pan. Ask students if their masses will be the same (no). Then ask if the amount of copper (volume) is the same for each object (no). Invite a discussion about

Chapter Pretest *continued*

whether or not the difference in mass and volume will change the fact that both of these items are made from the same substance: copper. Tell students that a substance has the same density, regardless of how much mass it has, because the volume will change in proportion to the mass. Be sure to emphasize Section 2, "Physical Properties," in Chapter 3, "Properties of Matter." Students must know density is mass per unit volume in order to master standard 8.8.a.

Question 8 *asks students to compute the density of a substance, given its mass and the information needed to calculate its volume.*

A Incorrect. This is the inverse of the density; students have computed volume over mass instead of mass over volume.

B Incorrect. Here, students have taken the reading off the second cylinder but did not subtract the initial volume of water.

C Correct. The volume (by water displacement) is 18 mL minus 12 mL, or 6.0 mL. 16.8 g divided by 6.0 mL equals 2.8 g/cm^3 (the density of granite).

D Incorrect. This is the volume of the rock.

Diagnostic Teaching Tip: Students who have difficulty answering this question correctly should be be shown how to compute the answer in a step-by-step process. Show students how the volumes recorded on the two cylinders should be read. Remind them that the two values must be subtracted to find the rock's volume. Then, compute the density given the values. Be sure to emphasize Section 2, "Physical Properties," in Chapter 3, "Properties of Matter." Students must know how to calculate the density of an irregular solid from measurements of mass and volume in order to master standard 8.8.b.

Question 9 *asks students to determine why a substance floats.*

A Incorrect. Color has nothing to do with determining a substance's density.

B Correct. Since corn oil is less dense than water, it will float on water.

C Incorrect. This is the freezing point of water but has nothing to do with why corn oil floats on water.

D Incorrect. Mass varies with size. Density is the determining factor.

Diagnostic Teaching Tip: Students who have difficulty with this question should view a demonstration. Show the class a small amount of corn oil in a glass. Then, slowly pour water into the glass. Explain that the corn oil floats because it is less dense than water. Tell students the same phenomenon occurs when ice cubes float in a glass of water. Be sure to emphasize Section 2, "Physical Properties," in Chapter 3, "Properties of Matter." Students must know how to predict whether an object will float or sink in order to master standard 8.8.d.

Chapter Pretest

_______ **1.** Which of the following measurements would NOT be accurate?
  A measuring the length of a paper clip using a centimeter ruler while looking at it from directly overhead
  B measuring 300 mL in a 500-mL measuring cup (increments every 100 mL) while looking at the line from eye level
  C measuring 500 g using a balance scale after checking it with a known mass
  D measuring 10 mL in a graduated cylinder while looking at the bottom of the meniscus at eye level

_______ **2.** You think your dog reacts differently when she sees you than when she sees other people. You decide to test your hypothesis. Which of the following is NOT a controlled parameter?
  A who enters the room
  B the number of people in the room
  C the noise level in the room
  D the time of day you run your tests

_______ **3.** Cissy decides to find out how much water is leaking from a pipe under the sink while her parents are buying repair material. She uses a graduated cylinder and takes measurements every 5 minutes. She collects the data below. Then she makes a graph, plotting time on the x-axis and total amount of water on the y-axis. Which of the following is true about the data?

Time (min)	5	10	15	20	25
Total amount of water (mL)	8	16	24	32	40

  A The slope of the line will be 8/5.
  B The slope of the line will be 8/40.
  C The slope of the line will be 25/5.
  D The slope of the line will be 25/40.

Chapter Pretest *continued*

______ **4.** A brass bell has a density of about 8.5 g/cm^3 and a mass of 170 g. Which of the following is the volume of the brass bell?
A 0.05 cm^3
B 0.5 cm^3
C 20 cm^3
D 1,445 cm^3

______ **5.** Vaughn's dad has been dieting for 5 weeks. As the weeks go by, he finds he is gradually losing less weight now than he did at the start of the diet. Which of the following would best describe a graph of his total weight loss over time?
A linear showing a direct relationship
B linear showing an inverse relationship
C nonlinear showing a direct relationship
D nonlinear showing an inverse relationship

______ **6.** When milk spoils, the molecules in the milk
A remain unchanged through time.
B move slower as they get older.
C form very fine milk crystals.
D interact to form a new product.

______ **7.** Which of the following can be said of the relationship between 1 g of gold and 1 kg of gold?
A The gram of gold is denser than the kilogram of gold.
B The kilogram of gold is denser than the gram of gold.
C Both pieces of gold have the same density.
D The volume of each piece of gold must be known to answer the question.

Chapter Pretest *continued*

_______ **8.** A rock has a mass of 16.8 g. It displaces water as shown below. Which of the following is its density?

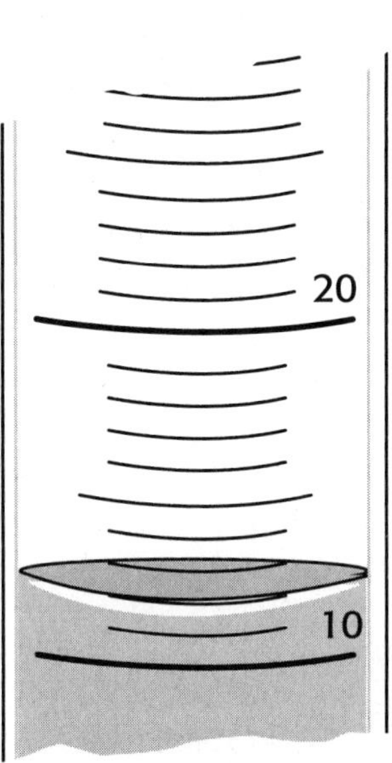
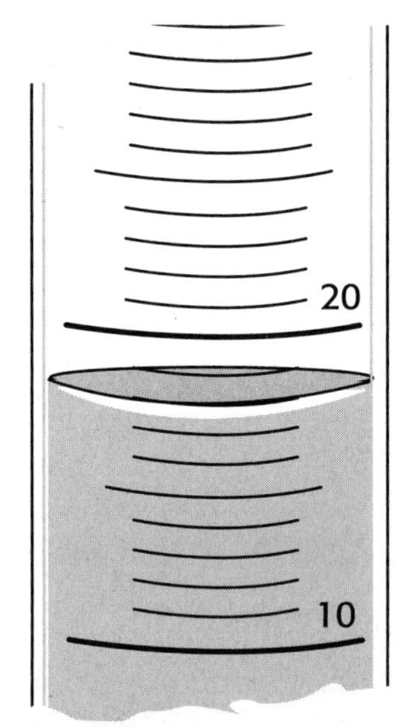

Before rock is added   **After rock is added**

A 0.36 g/cm^3
B 0.93 g/cm^3
C 2.8 g/cm^3
D 6.0 g/cm^3

_______ **9.** Why does corn oil float on water?
A Corn oil has color and water does not.
B Corn oil is less dense than water.
C Water freezes at 0°C.
D Water has more mass than corn oil.

Section Quiz

Section: What Is Matter?

Match the correct definition with the correct term. Write the letter in the space provided.

_______ **1.** a measure of the amount of matter in objects

_______ **2.** a measure of the gravitational force on objects

_______ **3.** the curve at a liquid's surface

_______ **4.** anything that has mass and takes up space

_______ **5.** the amount of space occupied by an object

a. volume

b. mass

c. matter

d. meniscus

e. weight

Write the letter of the correct answer in the space provided.

_______ **6.** What equation would you use to find the volume of a rectangular box?
　a. volume = length × width × height
　b. volume = length + width + height
　c. volume = length × width
　d. volume = length + width

_______ **7.** Which of the following units is best for expressing the volume of an irregular solid such as a rock?
　a. liters (L)
　b. cubic centimeters (cm^3)
　c. milliliters (mL)
　d. newtons (N)

_______ **8.** Which of the following statements is true about an object's weight but NOT about its mass?
　a. It may vary depending on the object's location.
　b. It is a measure of the amount of matter in the object.
　c. It is measured in kilograms (kg) or grams (g).
　d. It would be the same on the moon as it is on Earth.

_______ **9.** The SI unit of mass is the
　a. milliliter.
　b. cubic centimeter.
　c. kilogram.
　d. newton.

Name _________________________________ Class ______________ Date ____________

Section Quiz

Section: Physical Properties

Match the correct definition with the correct term. Write the letter in the space provided.

______ **1.** the ability to be made into thin sheets

______ **2.** a change in the form of a substance that does not change its identity

______ **3.** the ability to conduct electric current

______ **4.** the ratio of the mass of a substance to the volume of the substance

______ **5.** the ability of a substance to dissolve

______ **6.** characteristic of matter that can be observed without changing the matter's identity

a. density

b. physical property

c. solubility

d. physical change

e. malleability

f. electrical conductivity

Write the letter of the correct answer in the space provided.

______ **7.** Which of the following is the equation used to calculate a substance's density?

    a. $D = \dfrac{V}{m}$ **c.** $D = m + V$

    b. $D = \dfrac{m}{V}$ **d.** $D = V \times m$

______ **8.** What happens to a solid object with a density that is less than the density of water when it is placed in water?
    a. The object dissolves in the water.
    b. The object displaces a quantity of water greater than its volume.
    c. The object settles to the bottom of the water.
    d. The object floats on top of the water.

______ **9.** Which of the following is NOT an example of a purely physical change?
    a. the shaping of a gold bar   **c.** the explosion of fireworks
    b. the melting of a frozen fruit bar   **d.** the sanding of a piece of wood

______ **10.** What kinds of changes in substances are always physical changes?
    a. changes of state from solid to liquid to gas and back
    b. changes that result in new substances being formed
    c. changes that change the identity of the substances
    d. changes that change the density of the substances

Section Quiz

Section: Chemical Properties

Match the correct definition with the correct term. Write the letter in the space provided.

_______ **1.** the type of matter and its arrangement in an object

_______ **2.** the ability of a substance to burn

_______ **3.** the process of changing into entirely new substances

_______ **4.** a change in matter that does not change the identity of the substance

_______ **5.** the ability of a substance to change and form one or more new substances

a. reactivity

b. chemical change

c. flammability

d. composition

e. physical change

Write the letter of the correct answer in the space provided.

_______ **6.** Why are chemical properties harder to observe than physical properties?
  a. Chemical properties change substance's identity.
  b. Chemical properties depend on the size of the sample.
  c. Physical properties cannot be observed and measured.
  d. Physical properties change the identity of a substance.

_______ **7.** What is the best way to tell if a chemical change has taken place?
  a. The matter has changed color.
  b. The change is reversible.
  c. A mixture has separated into layers.
  d. The composition changes.

_______ **8.** Which of the following is NOT the result of a chemical change?
  a. soured milk
  b. rusted metal
  c. ground flour
  d. digested food

_______ **9.** What happens when most chemical changes take place?
  a. Only the form of the matter changes.
  b. The mass of the matter increases.
  c. There is no change in heat.
  d. Heat is liberated or absorbed.

Assessment

Chapter Test A

Properties of Matter
MULTIPLE CHOICE
Write the letter of the correct answer in the space provided.

______ **1.** What has mass and takes up space?
   a. weight
   b. volume
   c. space
   d. matter

______ **2.** How does a physical change differ from a chemical change?
   a. New properties of the matter are observed.
   b. New materials are produced.
   c. The change always involves heat or light.
   d. The identity of the matter is unchanged.

______ **3.** If you know an object's mass (m) and volume (V), what equation would you use to find its density?
   a. $D = m \times V$

   b. $D = \dfrac{V}{m}$

   c. $D = \dfrac{m}{V}$

   d. $D = V + m$

______ **4.** Which of the following signs does NOT necessarily indicate that a chemical change has happened?
   a. a change in state
   b. change in color or odor
   c. foaming or bubbling
   d. production of heat or light

______ **5.** What is any change in matter that changes only its physical form called?
   a. a formal change
   b. a physical change
   c. an identity change
   d. a chemical change

Chapter Test A *continued*

______ **6.** Which of the following usually happens during a chemical change?
  a. The identity of the matter stays the same.
  b. Only the state of the matter changes.
  c. Heat is released or absorbed.
  d. No heat is released or absorbed.

______ **7.** Which of the following units are best for expressing the density of a solid?
  a. g/mL
  b. m^3/kg
  c. g/cm^3
  d. N/cm^3

______ **8.** Why are chemical changes difficult to reverse?
  a. because they change the identity of the matter
  b. because they change the matter's form
  c. because they involve physical changes
  d. because their products are hard to find

MATCHING

Match the correct description with the correct term. Write the letter in the space provided.

______ **9.** aluminum made into thin sheets of foil

______ **10.** water in the form of solid ice

______ **11.** rust forming on metal

______ **12.** copper pulled into thin wires

a. ductility
b. reactivity
c. state of matter
d. malleability

Match the correct description with the correct term. Write the letter in the space provided.

______ **13.** flavored drink mix dissolving in water

______ **14.** objects floating or sinking in water

______ **15.** wood burning in a fireplace

______ **16.** a flower smelling sweet

a. density
b. solubility
c. odor
d. flammability

Chapter Test A *continued*

FILL-IN-THE-BLANK

Use the terms from the following list to complete the sentences below.

mass
property

volume
change

17. Things with _______________________ cannot share the same place at the same time.

18. A chemical _______________________ describes how a substance can take part in chemical reactions.

19. A chemical _______________________ is the process by which new substances are formed.

20. The amount of matter in an object is its _______________________.

Use the terms from the following list to complete the sentences below.

milliliters
kilogram

weight
meniscus

21. The SI unit for mass is the _______________________.

22. A measure of the gravitational force on an object is

the object's _______________________.

23. To measure volume with a graduated cylinder, look at

the _______________________.

24. You could use _______________________ to measure the volume of a soft-drink can.

Chapter Test B

Properties of Matter

MULTIPLE CHOICE

Write the letter of the correct answer in the space provided.

______ **1.** Which property of matter is a measure of gravitational force?
   a. density
   b. mass
   c. volume
   d. weight

______ **2.** If you know the volume (V) and mass (m) of an object, how do you find its density?
   a. Divide m by V.
   b. Multiply m by V.
   c. Divide V by m.
   d. Add m and V.

______ **3.** Which of the following is true of a chemical change?
   a. Only the state of the matter changes.
   b. The identity of the matter does not change.
   c. The identity of the matter changes.
   d. The change is easy to reverse.

______ **4.** Melting crayons is an example of a
   a. physical property.
   b. physical change.
   c. chemical property.
   d. chemical change.

______ **5.** Which of the following units would be best for describing the volume of vinegar used in an experiment?
   a. grams or kilograms
   b. meters or centimeters
   c. liters or milliliters
   d. newtons

______ **6.** Which of the following events is NOT a common sign that a chemical change has taken place?
   a. change in color or odor
   b. change in state
   c. foaming or bubbling
   d. production of heat or light

| Chapter Test B *continued*

_______ **7.** What chemical property accounts for iron rusting?
    a. flammability
    b. conductivity
    c. nonflammability
    d. reactivity with oxygen

_______ **8.** Why is a melting ice cube an example of a physical change?
    a. because the ice changes to another substance
    b. because the temperature changes
    c. because the form, not the identity, of the substance changes
    d. because the form and the identity of the substance changes

_______ **9.** What unit of density would be most appropriate for describing a solid bar of silver?
    a. g/mL
    b. g/cm^3
    c. oz/ft^3
    d. kg/L

_______ **10.** Which physical property of matter describes the relationship between mass and volume?
    a. density
    b. ductility
    c. reactivity
    d. weight

_______ **11.** Souring milk is an example of a
    a. physical property.
    b. physical change.
    c. chemical property.
    d. chemical change.

_______ **12.** Malleability is an example of a
    a. physical property.
    b. physical change.
    c. chemical property.
    d. chemical change.

_______ **13.** During most chemical changes, what is either liberated or absorbed?
    a. water
    b. oxygen
    c. heat
    d. mass

Chapter Test B *continued*

MATCHING

Match the correct description with the correct term. Write the letter in the space provided. Some terms will not be used.

_______ **14.** The saltiness of sea water is the result of this property.

_______ **15.** Whether objects float or sink depends on this property.

_______ **16.** This is the physical form in which a substance exists.

_______ **17.** This is the type and arrangement of matter that makes up an object.

_______ **18.** This is a measure of the amount of matter in a object.

_______ **19.** This is the ability of a substance to resist burning.

_______ **20.** This is anything that has mass and takes up space.

_______ **21.** This is the rate at which a substance conducts heat.

a. thermal conductivity

b. composition

c. nonflammability

d. matter

e. state of matter

f. solubility

g. reactivity

h. mass

i. ductility

j. density

Chapter Test B *continued*

MULTIPLE CHOICE

The table below shows the density of some common substances. Use the table to answer questions 22 through 25. Write the letter of the correct answer in the space provided.

SUBSTANCE	DENSITY (g/m³)
Aluminum (solid)	2.7
Ice (solid)	5.02
Pyrite (solid)	13.55
Mercury (liquid)	13.55
Silver (solid)	10.50
Ice (solid)	0.93
Water (liquid)	1.00
Zinc (solid)	7.13
Wood (oak)	0.85

_______ **22.** A cube has a density of 2.7 g/cm^3. What substance is the cube made of?
  a. aluminum
  b. ice
  c. pyrite
  d. wood

_______ **23.** What substance has a density more than 13 times greater than water?
  a. ice
  b. silver
  c. aluminum
  d. mercury

_______ **24.** Why does ice float on top of liquid water?
  a. Ice has a lower density than water.
  b. Ice has a higher density than water.
  c. Ice is a solid.
  d. Ice is colder than water.

_______ **25.** If you have a sample of a solid that has a volume of 2 cm^3 and a mass of 21 g, what substance is it?
  a. mercury
  b. silver
  c. zinc
  d. pyrite

Chapter Test C

Properties of Matter

USING KEY TERMS

Use the terms from the following list to complete the sentences below. Each term may be used only once. Some terms may not be used.

weight	mass	density
volume	physical change	meniscus
chemical change	chemical	physical

1. When you measure the volume of liquids, you must read the amount at the

bottom of the _____________________.

2. The physical property of matter that describes the relationship between

mass and volume is _____________________.

3. Water evaporating from a puddle is an example of a(n)

_____________________.

4. One way to learn about the _____________________ properties of a(n)
substance is to observe what new substances form during a reaction.

5. An object's _____________________ is affected by the gravitational force
on the object.

6. A copper penny can turn green if it reacts with carbon dioxide and water.

This is an example of a(n) _____________________.

UNDERSTANDING KEY IDEAS

Write the letter of the correct answer in the space provided.

______ **7.** The amount of a liquid used in scientific experiments is typically
expressed in
    a. centimeters or meters.   **c.** liters or milliliters.
    b. grams or milligrams.   **d.** ounces or gallons.

______ **8.** Which of the following is NOT a physical property of matter?
    a. ductility   **c.** thermal conductivity
    b. color   **d.** reactivity with water

______ **9.** If you have 7 mL of a liquid that has a mass 8.82 g, what is the density
of the liquid?
    a. 0.79 mL/g   **c.** 1.26 mL/g
    b. 0.79 g/mL   **d.** 1.26 g/mL

Chapter Test C *continued*

______ **10.** During physical changes, matter always retains its
     a. size.   **c.** state.
     b. identity.   **d.** texture.

______ **11.** To compare the densities of oil and water, pour the liquids into a container and observe how they
     a. change color.   **c.** separate into layers.
     b. evaporate quickly.   **d.** create an odor.

12. Explain why volume and mass are not characteristic properties of matter.

____

____

____

____

13. Describe two ways to compare the densities of unknown substances. Why is density a useful property for identifying matter?

____

____

____

____

14. Why could an astronaut carry more massive equipment around on the moon than on Earth?

____

____

____

____

____

15. Describe the part that heat plays in most chemical changes.

____

____

____

____

____

Chapter Test C *continued*

CRITICAL THINKING

16. After a tree is cut with a chain saw, it is impossible to put the tiny wood chips back together. The process cannot be reversed. Does this mean that cutting trees with a chain saw causes a chemical change in the wood? Explain why or why not.

____

____

____

____

____

____

17. Summarize the differences between mass and weight. Why do you think people tend to confuse these terms?

____

____

____

____

____

18. A glass cylinder contains four liquids in four separate layers. One liquid is pure water. The purple liquid has a density 1.62 g/cm^3. The yellow liquid has a density of 0.46 g/cm^3. The red has a density of 0.91 g/cm^3. What is the order of the liquids in the cylinder? Explain your answer. What will happen if you slip a small, flat chip of wood (density 0.85 g/cm^3) into the cylinder?

____

____

____

____

____

____

CONCEPT MAPPING

19. Use the following terms to complete the concept map below:

    identity   chemical changes   burning
    dissolving   substances   physical changes

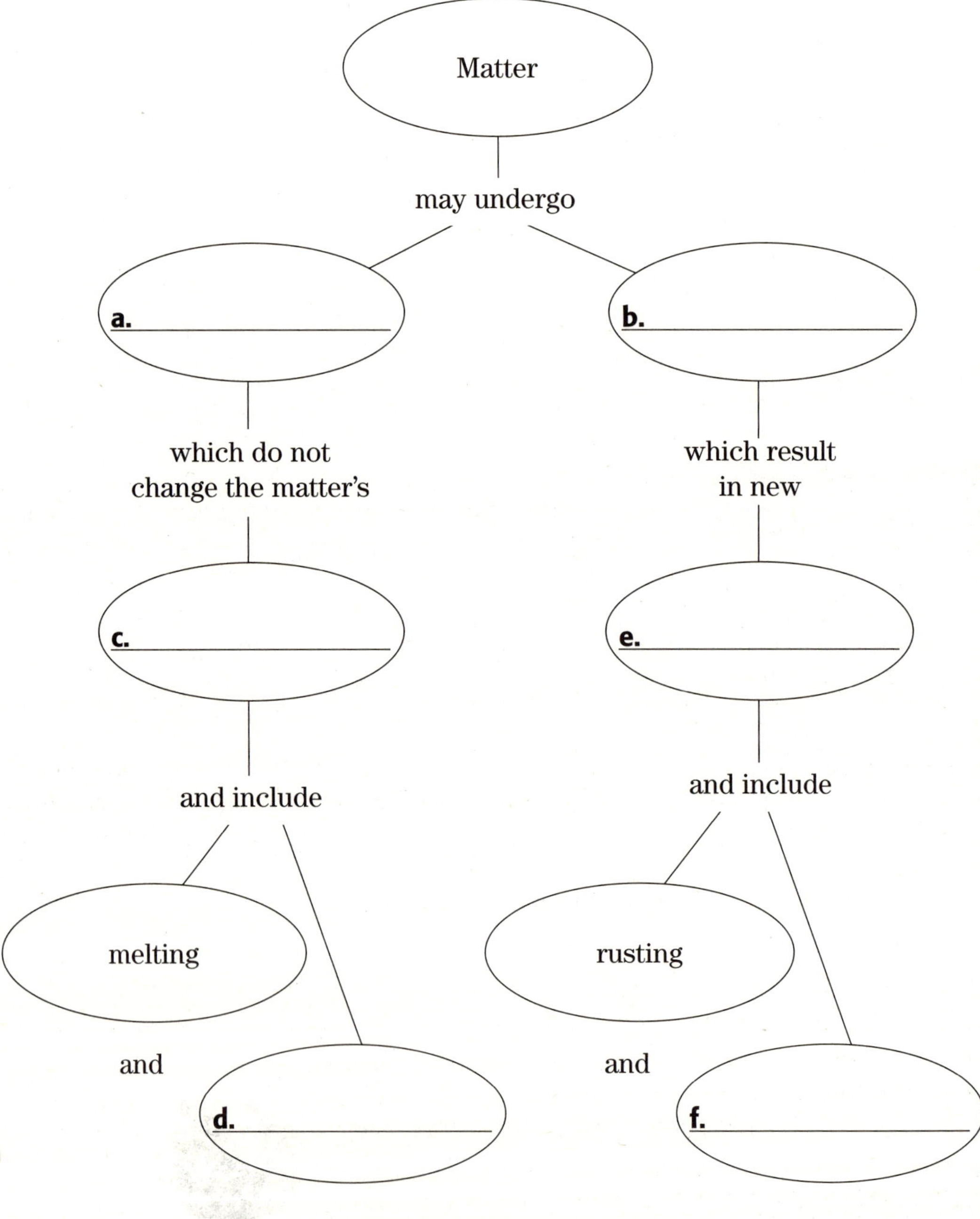

Assessment

Performance-Based Assessment

Teacher Notes

PURPOSE

Students will measure the mass and volume of copper and zinc to calculate their densities. They will use this information to determine which coins are made of copper and which are made of zinc.

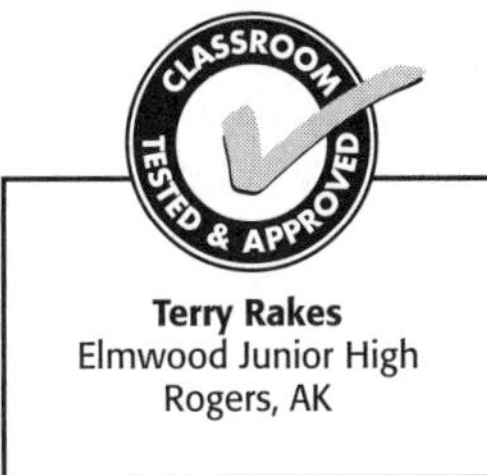

Terry Rakes
Elmwood Junior High
Rogers, AK

TIME REQUIRED

One 45-minute class period. Students will need 30 minutes at the activity station and 15 minutes to answer the analysis questions.

RATING

Easy ← 1　2　3　4 → Hard

Teacher Prep–2
Student Set-up–1
Concept Level–2
Clean Up–1

ADVANCE PREPARATION

Make sure you use graduated cylinders that measure in 1.0 mL increments. The diameter of the cylinder should be wide enough so that the largest coin that you will test can easily fall in and out. (You don't want it to get stuck!) Cover each activity station with a tarp or a drop cloth. Equip each activity station with the necessary materials.

SAFETY INFORMATION

Spilled water is a slipping hazard. Clean up water spills immediately. Students should immediately notify the teacher if a graduated cylinder breaks or if a student cuts himself or herself. Have a sharps container nearby in case of glass breakage.

TEACHING STRATEGIES

This activity works best in groups of 4–5 students. Before the activity, review how to calculate density. Show students how to correctly read a water level in a graduated cylinder. Remind students that in order to get an accurate reading, they must read the amount at the bottom of the meniscus.

Pennies minted before 1982 were made of 95 percent copper and 5 percent zinc. From 1982 and thereafter, pennies were made of copper-plated zinc. Nickels are 75 percent copper and 25 percent nickel. Dimes and quarters are manufactured with a pure copper core and a face that is 75 percent copper and 25 percent nickel.

Although none of the coins will have precisely the same density as the pure metals, students should be able to distinguish whether the coins consist mostly of copper or zinc.

Performance-Based Assessment *continued*

The pennies used in this activity should all be minted either before or after 1982. This will ensure consistency in the measurement of density.

Evaluation Strategies

Use the following rubric to help evaluate student performance.

Rubric for Assessment

Possible points	Appropriate use of materials and equipment (30 points possible)
30–20	Successful completion of activity; safe and careful handling of materials and equipment; attention to detail; superior lab skills
19–10	Activity is generally complete; successful use of materials and equipment; sound knowledge of lab techniques; somewhat unfocused performance; mild neglect of safety measures
9–1	Attempts to complete activity yield inadequate results; unsafe lab technique; apparent lack of skill
	Making observations and testing hypotheses (40 points possible)
40–27	Superior analysis stated clearly and accurately; high level of detail; correct usage of scientific terminology
26–14	Complete analysis; moderate level of detail, but expressed in unclear manner; minor inaccuracies; some errors or inconsistencies
13–1	Erroneous, incomplete, or unclear analysis; complete lack of accuracy and detail
	Drawing conclusions (30 points possible)
30–20	Clear, detailed explanation shows good understanding of density; use of examples support explanations
19–10	Adequate understanding of density with minor difficulty in expression
9–1	Poor understanding of density; explanation unclear or not relevant to the activity; substantial factual errors

Performance-Based Assessment

OBJECTIVE

Every metal has a unique density. You will measure these densities to find out what metals coins are made of.

KNOW THE SCORE!

As you work through the activity, keep in mind that you will be earning a grade for the following:

- how well you work with the materials and equipment (30%)
- how well you make observations and test the hypothesis (40%)
- how accurately you identify test objects (30%)

ASK A QUESTION

How can I determine which metal I have?

MATERIALS AND EQUIPMENT

- balance
- calculator
- copper sample
- cylinder, graduated (1.0 mL increments)
- dimes (10)

- nickels (10)
- pennies (10)
- quarters (5)
- water, 1 L
- zinc sample

SAFETY INFORMATION 

- Clean up water spills immediately; spilled water is a slipping hazard.

Using Scientific Methods

MAKE OBSERVATIONS

1. Compare the luster and appearance of the coins with the luster and appearance of the copper and zinc samples.

____

____

____

Performance-Based Assessment *continued*

FORM A HYPOTHESIS

2. Which metals do you think the coins are mostly made of? Support your hypothesis with observations.

____

____

____

____

____

TEST THE HYPOTHESIS

3. Place the copper sample on the balance. Measure and record the mass in the table.

4. Pour water into the graduated cylinder until it is half full. Record the starting volume of the water in the table.

5. Place the copper sample in the cylinder. Record the ending volume of the water in the table.

6. Subtract the starting volume from the ending volume to find the volume of the metal. Record this value in the table below.

7. Repeat steps 3–6 with the zinc sample. Record your measurements.

8. Repeat steps 3–6 with the 10 pennies. Record your measurements. Continue to repeat steps 3–6 until all the sets of coins have been tested.

ANALYZE THE RESULTS

9. Calculate the density of each metal by dividing the mass of the metal by the volume of the metal. Record the density in the table on the next page. Show your work below.

Performance-Based Assessment *continued*

MATERIALS DATA

Metal sample	Starting volume (mL)	Ending volume (mL)	Volume of metal (mL)	Mass of metal (g)	Density of metal (g/mL)
Copper					
Zinc					
Penny					
Nickel					
Dime					

BIG IDEA QUESTION

10. Draw Conclusions Compare the densities of the coins with the densities of the known metals. What are the coins mostly made of? Support your answer.

____

____

____

____

____

____

Standards Assessment

Teacher Notes and Answer Key

To provide practice under more realistic testing conditions, give students 20 min to answer all of the questions in this assessment.

QUESTION NUMBER	CORRECT ANSWER	STANDARD
1	*C*	8.8.a (supporting)
2	*B*	8.5.c (supporting)
3	*A*	8.5.a (supporting)
4	*D*	8.5.d (supporting)
5	*B*	8.8.a (mastering)
6	*D*	8.8.b (supporting)
7	*D*	8.7.c (mastering)
8	*B*	8.5.c (mastering)
9	*B*	8.8.b (mastering)
10	*D*	8.5.a (mastering)
11	*A*	8.8.d (mastering)
12	*A*	5.1.c (mastering)
13	*B*	5.1.f (mastering)
14	*D*	5.1.i (mastering)
15	*C*	5.1.d (mastering)

TEST DOCTOR

The following Standards Assessment questions have been diagnosed by the Test Doctor. Find out what might be causing your students' "ailing" answers. Each Test Doctor is followed by a diagnostic teaching tip to help you address students' learning needs.

Question 1 *asks students to define the meaning of the word* volume.

A Incorrect. While *size* and *amount* are similar in meaning, *amount* refers specifically to quantity.

B Incorrect. *Density* is the relationship between the mass of an object and the volume of an object.

C Correct. *Volume*, the amount of matter in a three-dimensional space, is the scientific term for *size*.

D Incorrect. *Mass* is the amount of matter an object or substance contains.

Diagnostic Teaching Tip: Students who have difficulty answering this question correctly might benefit from reviewing the definition of volume. Instruct students to draw several three-dimensional shapes according to specific measurements—make sure at least two are different sizes of the same shape. Have them calculate the volume of those shapes. Then have them write several sentences describing what it means to calculate volume. What information does volume communicate? Did the size of the object affect the volume?

Standards Assessment *continued*

Question 2 *asks students to define the word* liberate *in context.*

A **Incorrect.** *Steals,* which means "takes without permission," does not fit the context of the sentence.

B **Correct.** *Liberates* means *frees.* The process of burning wood frees, or liberates, heat.

C **Incorrect.** *Creates* means "causes to come into existence." *Creates* does not work in this sentence because heat is released by the burning process, not brought into being.

D **Incorrect.** *Imitates,* which means "to be or become like in appearance," does not fit the context of the sentence.

Diagnostic Teaching Tip: Students who struggle to answer this question correctly might benefit from exploring the meanings of *liberate* in the dictionary. Have students write out the definition for liberate and then look up, and write down, synonyms for *liberate* in a thesaurus.

Question 3 *asks students to define the term* chemical reaction.

A **Correct.** A chemical reaction is a process that changes the molecular composition of a substance.

B **Incorrect.** While burning something does cause a chemical reaction, igniting something and burning it is a process and not the definition of a chemical reaction.

C **Incorrect.** A chemical reaction affects *both* appearance and molecular structure.

D **Incorrect.** Chemical reaction does not necessarily involve using chemicals. Some chemicals do cause chemical reactions, however.

Diagnostic Teaching Tip: Students who have difficulty answering this question correctly might benefit from practice distinguishing between physical changes and changes occurring as a result of chemical reactions. Ask students to discuss how water changes states from solid to liquid to gas. Does the molecular structure of the substance change as it changes states? Is it still water when it is ice? Then, ask students to discuss what happens to a piece of wood when in burns. After it has burned, is it still wood? Ask students to analyze the differences between the two types of changes.

Question 4 *asks students define the word* physical.

A **Incorrect.** Chemicals are not necessary for determining density.

B **Incorrect.** Density is the ratio of the mass of a substance to the *volume* of the substance; however, that fact does not provide a definition for the term *physical.*

C **Incorrect.** While *physical* can be used when discussing the body, *density* is not a term commonly used to refer to the body; therefore, it does fit the context of the sentence.

D **Correct.** A *physical* property is a characteristic of matter that does not involve a chemical change, such as density, color, or hardness.

Standards Assessment *continued*

Diagnostic Teaching Tip: Students who have difficulty answering this question correctly might benefit from reviewing the properties of matter. Choose a few random objects and play "Twenty Questions." How did knowing the properties of matter help in identifying the objects?

Question 5 *asks students to demonstrate understanding that density is mass per unit volume.*

A Incorrect. *Weight* is determined by measuring the gravitational force acting on an object.

B Correct. *Density* can be determined by dividing the mass of an object by its volume.

C Incorrect. *Ductility* is the ability of a substance to be pulled into wire.

D Incorrect. *State* is the physical form in which a substance exists, such as a solid, liquid, or gas.

Diagnostic Teaching Tip: Students who struggle to answer this type of question might benefit from further practice calculating the density of objects. Gather several objects that will sink in water. Have students create a graphic organizer for their data that includes columns for recording mass, volume, and density. Help students measure the mass and volume of the objects. Have them calculate the density of the objects. After they have gathered their data, discuss and compare the results.

Question 6 *asks students to demonstrate understanding of how to determine the volume of a regularly shaped solid.*

A Incorrect. 12 cm^2 correctly expresses area, not volume. Area can be calculated by multiplying length times width.

B Incorrect. This answer has not correctly calculated length $\times$ width $\times$ height.

C Incorrect. This answer incorrectly expresses area, not volume.

D Correct. Volume is determined by calculating area (A) $\times$ height (H). The area of the base of this shape is 12 cm^2. Its height is 3 cm. $12 \text{ cm}^2 \times 3 \text{ cm} = 36 \text{ cm}^3$. The volume of the solid is 36 cm^3.

Diagnostic Teaching Tip: Students who have difficulty answering this question correctly might benefit from a review of the differences between area and volume and the dimensions these measurements occupy. Have students draw a square and a cube. Exchange papers with a neighbor. Ask them to determine the area of the square, and then write a sentence about how many dimensions they used to find the area. Remind them that the superscript number 2 on their answer indicates the 2 dimensions they used to determine the area. Repeat this process for the three-dimensional shape.

Question 7 *asks students to demonstrate understanding of the difference between physical and chemical properties.*

A Incorrect. While copper is *conductive*, conductivity is a physical property rather than a chemical one.

Standards Assessment *continued*

B Incorrect. While copper is *ductile*, ductility is a physical property rather than a chemical one.

C Incorrect. Copper is not *soluble*, and solubility is a physical property rather than a chemical one.

D Correct. Copper is *reactive* with air. This chemical reaction forms a green copper compound, which is why the Statue of Liberty looks green rather than reddish brown, the color of copper.

Diagnostic Teaching Tip: Students who have difficulty answering this question correctly might benefit from reviewing the definitions of physical and chemical properties. Create two tables listing the same selection of objects and substances you have gathered. One table should have a column for physical properties and a column for physical changes, while the other table should have one column for chemical properties and one column for chemical changes. First have students describe the physical properties of the objects and substances. Then ask them to describe any chemical properties they can think of. Select an object or substance and describe a change that might occur. Have students determine if that change would be physical or chemical. Record their data in the tables.

Question 8 *asks students to demonstrate understanding that chemical changes often liberate or absorb heat.*

A Incorrect. When a substance changes shape or state, it is changing physically, not chemically.

B Correct. Chemical changes usually liberate or absorb heat. Other signs that a chemical change is occurring include a change in odor or color, and fizzing or foaming.

C Incorrect. Density and malleability are physical properties. They do not indicate any type of change, physical or chemical.

D Incorrect. Flammability and reactivity are chemical properties; however, they indicate only that a certain type of chemical reaction *could* take place under certain conditions, not that it *is* taking place.

Diagnostic Teaching Tip: Students who choose answer A or C may be confusing chemical changes and properties with physical changes and properties. These students might benefit from creating a poster that illustrates the differences between these concepts. Students who choose answer C or D may be confusing properties and changes. These students might benefit from creating a lab that tests the differences between these two concepts. Have students conduct the labs they have created and share the results with the class.

Question 9 *asks students to demonstrate how to calculate the density of substances from mass and volume.*

A Incorrect. The solid sample, which has a density of 0.75 g/ cm^3, is less dense than mercury, which has a density of 13.55 g/cm^3.

B Correct. The solid sample has a density of 0.75 g/cm^3, while water has a density of 1.00 g/cm^3. The rock sample is less dense than water.

Standards Assessment *continued*

C Incorrect. The solid sample, which has a density of 0.75 g/ cm^3, is denser than helium, which has a density of 0.0001663 g/cm^3.

D Incorrect. The solid sample, which has a density of 0.75 g/ cm^3, is less dense than zinc, which has a density of 7.13 g/cm^3.

Diagnostic Teaching Tip: Students who have difficulty answering this question correctly might benefit from practice determining the density of objects in relation to other objects. For example, you might have students determine the density of several objects and graph their results. Then present a few more objects and ask students to predict their density in relation to the objects on the graph. Allow them to handle the objects before making their predictions. Tally students' predictions on the board. Have students measure the density of the objects. Discuss how the results compare to their predictions. Have students plot the densities of the final objects on the graph.

Question 10 *asks students to demonstrate understanding of the differences between physical and chemical changes.*

A Incorrect. The opposite is true: A chemical change is often difficult to undo, while a physical change is often easy to undo.

B Incorrect. The opposite is true: Physical changes are more easily observed than chemical changes.

C Incorrect. The opposite is true: A chemical change affects the molecular structure of a substance, while a physical change affects only its physical properties.

D Correct. A physical change merely alters the physical properties of a substance, such as the way water looks when it freezes. A chemical change alters the molecular structure of a substance, such as dissolving an effervescent tablet in water.

Diagnostic Teaching Tip: Students who have difficulty answering this question correctly might benefit from conducting short experiments in which they can observe some of the differences between these two types of changes. For example, have students crack, scramble, and cook eggs. Have them keep a record of the changes that take place throughout the process. Then have students mix small amounts of baking soda with vinegar and record the changes they observe. Discuss the results of these two experiments and ask students to classify the changes they observed as physical or chemical.

Question 11 *asks students to demonstrate understanding that whether an object will float or sink is predictable.*

A Correct. If the *density* of an object is less than the density of water, the object will float. Likewise, a solid object whose density is greater than the density of water will sink.

B Incorrect. *Volume* is used to determine an object's density ($D = m/V$).

C Incorrect. While *malleability*, the ability of a substance to be rolled or pounded into thin sheets, is a physical property, it is not used to determine if a substance will float.

D Incorrect. While *conductivity*, the rate at which a substance transfers heat and electricity, is a physical property, it is not used to determine if a substance will float.

Diagnostic Teaching Tip: Students who have difficulty answering this question correctly might benefit from reviewing density. Choose several solids. Pass one of the solids around the room, and then have the students predict whether it will float or sink. Record the prediction. Choose a student to measure its mass and volume. Choose another student to calculate density. Record the calculation. Place the object in a container of water. Record the result. Discuss whether the prediction, calculation and outcome match. Repeat for the remaining objects.

Question 12 *asks students to identify common properties of elemental metals.*

A Correct. All metals conduct heat and electricity better than nonmetals. Elemental metals like these have the physical property of high thermal and electrical conductivity.

B Incorrect. One physical property of metals is that they are malleable.

C Incorrect. *Reactivity* varies widely among metals, so this statement is not true.

D Incorrect. Some metals, like mercury, are liquid at room temperature; however, none of the metals listed are liquid at room temperature.

Diagnostic Teaching Tip: Students who have difficulty answering this question correctly might benefit from reviewing the properties of metals. Have students brainstorm the different physical and chemical properties of metals. Make a mixed selection of metals and nonmetals available to students and have them conduct experiments to determine which substances are metals and which are nonmetals.

Question 13 *asks students to demonstrate knowledge of the differences between mixtures and compounds.*

A Incorrect. The test the student is conducting is not to determine if the substance is metallic or nonmetallic.

B Correct. The liquid solution is a mixture because it is possible to separate the large particles and the small particles. Physical processes can separate the constituent parts of mixtures.

C Incorrect. The test the student is conducting is not to determine if the substance is metallic or nonmetallic.

D Incorrect. The material is not a compound. The constituent parts of compounds cannot be separated by physical means.

Standards Assessment *continued*

Diagnostic Teaching Tip: Students who have difficulty answering this question correctly might benefit from comparing mixtures and compounds. Create a set of instructions for making a mixture and another set of instructions for making a compound, without indicating which is which. Have students follow each set of instructions, recording their data and observations as they go. Afterward, ask students to share their data and observations. Can they tell which substance is a compound and which is a mixture? If not, brainstorm ways of determining which is which. Have students select the best method and use it to classify each solution.

Question 14 *asks students to demonstrate understanding of the common properties of elements.*

A Incorrect. Chlorine (Cl) is a gas, but sodium (Na) is a solid.

B Incorrect. Sodium (Na) is a solid, but chlorine (Cl) is a gas.

C Incorrect. Sodium (Na) is a metal, but chlorine (Cl) is not.

D Correct. Salt is a compound made from sodium (Na), a metallic element, and chlorine (Cl), a nonmetallic element.

Diagnostic Teaching Tip: Students who have difficulty answering this question correctly might benefit from practice identifying the properties of compounds. Provide a list of common compounds, such as water, sugar, and carbon dioxide, and their atomic structures. Have students describe the properties of each element in a compound, and then analyze how the properties of the elements change when they react with other elements to form compounds.

Question 15 *asks students to demonstrate understanding that each element is made of one kind of atom.*

A Incorrect. An element is not made of two kinds of atoms; a molecule usually consists of two or more atoms joined in a definite ratio.

B Incorrect. An element is not made of molecules; in fact, most molecules are composed of atoms of two or more elements.

C Correct. An element is a substance made of one kind of atom. Repeated experiments have shown that elements cannot be reduced to more basic substances.

D Incorrect. Atoms are not made of molecules; molecules are made of atoms.

Diagnostic Teaching Tip: Students who have difficulty answering this question correctly might benefit from practice making models of simple elements. Assign students different elements and help them create three-dimensional models of one atom of that element. Have them attach other identical atoms to the original atom to make an element. Ask them to write a short description of the structure of their elements. Display the elements and descriptions in the classroom.

Name _______________________________ Class _________________ Date _____________

Standards Assessment

REVIEWING ACADEMIC VOCABULARY

_______ **1.** Which of the following words is the closest in meaning to "size"?
   A amount
   B density
   C volume
   D mass

_______ **2.** In the sentence "The process of burning wood liberates heat," what does the word *liberates* mean?
   A steals
   B releases
   C creates
   D imitates

_______ **3.** In the sentence "Ash is formed by the chemical reaction between wood and fire," what does the term *chemical reaction* mean?
   A a process that changes the molecular composition of a substance
   B a process of igniting something, burning it, and recording the results
   C a process that affects the appearance but not the molecular structure
   D a process of using chemicals to break down molecular structures

_______ **4.** In the sentence "Density is a physical property," what does the term *physical* mean?
   A having to do with chemicals
   B having to do with volume
   C having to do with the body
   D having to do with matter

REVIEWING CONCEPTS

_______ **5.** Which physical property of an object can be determined by dividing its mass by its volume?
   A weight
   B density
   C ductility
   D state

Standards Assessment *continued*

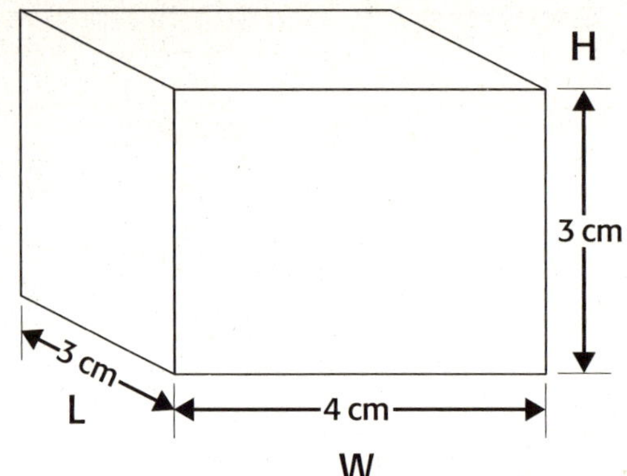

______ **6.** What is the volume of the solid pictured above?
 A 12 cm^2
 B 12 cm^3
 C 36 cm^2
 D 36 cm^3

______ **7.** Which of the following is a chemical property that describes copper?
 A conductive
 B ductile
 C soluble
 D reactive

______ **8.** Which of the following describes signs that a chemical change is occurring?
 A A substance changes shape or state.
 B A substance gives off or absorbs heat.
 C A substance is dense and malleable.
 D A substance is flammable and reactive.

Standards Assessment *continued*

Densities of Common Substances*			
Substance	**Density* (g/cm³)**	**Substance**	**Density* (g/cm³)**
Helium (gas)	0.0001663	Zinc (solid)	7.13
Oxygen (gas)	0.001331	Silver (solid)	10.50
Water (liquid)	1.00	Lead (solid)	11.35
Pyrite (solid)	5.02	Mercury (liquid)	13.55

*at 20°C and 1.0 atm

______ **9.** A solid sample has a mass of 22.5 g and displaces 30 mL of water. Use the table above to determine which sentence best describes the solid sample.
  A It is denser than mercury.
  B It is less dense than water.
  C It is less dense than helium.
  D It is denser than zinc.

______ **10.** Substances can undergo physical changes or chemical changes. What is the difference between these two kinds of changes?
  A A chemical change can often be undone, and a physical change cannot.
  B A physical change cannot be observed easily, and a chemical change can.
  C A chemical change affects only the physical properties of a substance. A physical change changes the molecular structure of a substance.
  D A physical change does not affect the identity of a substance. A chemical change changes the molecular structure of a substance.

______ **11.** Which physical property can you use to determine if a substance will float in water?
  A density
  B volume
  C malleability
  D conductivity

| Standards Assessment *continued*

REVIEWING PRIOR LEARNING

______ **12.** What properties do the metals aluminum, copper, silver, and gold have in common?
  A They conduct heat and electricity well.
  B They are brittle and do not bend easily.
  C They do not chemically react.
  D They are liquid at room temperature.

______ **13.** A student pours a material through a filter. Some particles get caught in the filter and some pass through. This material can be best described as a
  A nonmetal.
  B mixture.
  C metal.
  D compound.

______ **14.** Which of the following sentences best describes table salt, NaCl?
  A Table salt is a compound made from two gases.
  B Table salt is a compound made from a solid and a liquid.
  C Table salt is a compound made from two metals.
  D Table salt is a compound made from a metal and a nonmetal.

______ **15.** An element is made up of
  A two kinds of atoms.
  B one kind of molecule.
  C one kind of atom.
  D two kinds of molecules.

DATASHEET

Similar Size, Different Mass

Teacher Notes

In this activity, students learn that two objects can have the same size yet have different masses (covers standard 8.7.c). If you have a double-pan balance, you can use it to quickly compare the mass of each ball to an object of similar size. For example, show students that a smaller ball of steel (or another dense metal) easily counterbalances the golf ball.

MATERIALS

For each group

- balance, triple-beam
- golf ball
- table-tennis ball

Similar Size, Different Mass

In this activity, you will compare the physical properties of two objects that are similar in size and shape.

PROCEDURE

1. Examine a **table-tennis ball.**
   - Then, examine a golf ball.

2. Use a **triple-beam balance** to determine the mass of each ball.
   - Write down the mass of the table-tennis ball: ____________________
   - Write down the mass of the golf ball: ____________________

3. Compare the volume of the balls. (Hint: Are the balls the same size or is one bigger?)

____

____

   - Look at each ball for five minutes. List the properties you can identify. (Hint: List everything you can see and say about the balls.)
   - When I look at the table-tennis ball, I see:

____

____

   - When I look at the golf ball, I see:

____

____

____

ANALYSIS

4. At the end of five minutes, discuss your findings with the class.

5. What units did you use to express the mass of each ball? (Hint: See step 2.)

____

   - Think about the amount of matter in each ball.
   - Is the amount of matter the same?

____

   - If not, which ball has more matter?

____

Similar Size, Different Mass *continued*

6. Think about two objects that are similar in size.
   • Does that mean the two objects have the same amount of matter?

____

   • Explain your answer.

____

____

Explore Activity   **DATASHEET B**

Similar Size, Different Mass

In this activity, you will compare the physical properties of two objects that are similar in size and shape.

PROCEDURE

1. Examine a **table-tennis ball** and a **golf ball.**

2. Use a **triple-beam balance** to determine the mass of each ball. Record your observations.

3. Compare the volume of the balls. For each ball, identify properties other than mass and volume by observing the ball for five minutes. List the properties that you identify.

ANALYSIS

4. Five minutes after observing the two balls, discuss your findings with the class.

5. What units did you use to express the mass of each ball? What can you conclude about the amount of matter in each ball?

____

____

____

____

6. For any two objects that are similar in size, are both objects made up of the same amount of matter? Explain your answer.

____

____

____

____

____

____

Similar Size, Different Mass

In this activity, you will compare the physical properties of two objects that are similar in size and shape.

PROCEDURE

1. Examine a **table-tennis ball** and a **golf ball.**

2. Use a **triple-beam balance** to determine the mass of each ball. Record your observations.

____

____

3. Compare the volume of the balls. For each ball, identify properties other than mass and volume by observing the ball for five minutes. List the properties that you identify.

____

____

____

____

ANALYSIS

4. Five minutes after observing the two balls, discuss your findings with the class.

5. What units did you use to express the mass of each ball? What can you conclude about the amount of matter in each ball?

____

____

____

____

Similar Size, Different Mass *continued*

6. For any two objects that are similar in size, are both objects made up of the
same amount of matter? Explain your answer. Then, list four more sets of
objects that are similar in size but are different in mass.

____

____

____

____

____

____

____

DATASHEET

Finding Accurate Volumes

Teacher Notes

In this activity, students will learn why reading a graduated cylinder at eye level is important for measuring the volume of a liquid accurately (covers 8.8.b and 8.9.b).

MATERIALS

For each group

- graduated cylinder, 100 mL
- water, colored with food coloring

SAFETY CAUTION

Remind students to review all safety cautions and icons before beginning this activity.

(Quick Lab)   **DATASHEET A**

Finding Accurate Volumes

SAFETY INFORMATION

PROCEDURE

1. Pour **colored water** into a **100 mL graduated cylinder** up to the 50 mL mark.
   - Be sure to look at eye level when you read the volume.
   - Make sure to measure the volume from the bottom of the meniscus. (Hint: The meniscus is the curve at a liquid's surface.)

2. Look at the meniscus from different positions as described below.

   a. Look at the meniscus with your head slightly above the graduated cylinder (above eye level). Read the volume of water and write down your reading here. _______________________

   b. Look at the meniscus with your head even further above the graduated cylinder (above eye level).Read the volume of water and write down your reading here. _______________________

   c. Look at the meniscus with your head slightly below the graduated cylinder (below eye level). Read the volume of water and write down your reading here. _______________________

   d. Look at the meniscus with your head even further below the graduated cylinder (below eye level). Read the volume of water and write down your reading here. _______________________

3. Compare your measurements from step 2 to your measurement at eye level (50 mL).

   a. Note any difference between the measurement from step 2a and 50 mL.

   ____

   b. Note any difference between the measurement from step 2b and 50 mL.

   ____

   c. Note any difference between the measurement from step 2c and 50 mL.

   ____

   d. Note any difference between the measurement from step 2d and 50 mL.

   ____

4. Why do you think reading a meniscus at eye level is important for measuring volume?

   ____

Quick Lab

Finding Accurate Volumes

SAFETY INFORMATION

PROCEDURE

1. Pour **colored water** into a **100 mL graduated cylinder** up to the 50 mL mark. Measure from the bottom of the meniscus as seen at eye level.

2. Look at the meniscus from two positions above eye level and two positions below eye level. At each position, read the volume of water. Record your readings.

3. Compare your measurement at each position to your eye-level measurement (50 mL). Describe any differences between the measurements.

____

____

____

____

____

4. Why is reading a meniscus at eye level important for measuring volume accurately?

____

____

____

____

____

DATASHEET C

Finding Accurate Volumes

SAFETY INFORMATION

PROCEDURE

1. Pour **colored water** into a **100 mL graduated cylinder** up to the 50 mL mark. Measure from the bottom of the meniscus as seen at eye level.

2. Look at the meniscus from two positions above eye level and two positions below eye level. At each position, read the volume of water. Record your readings.

____

____

____

____

3. Compare your measurement at each position to your eye-level measurement (50 mL). Describe any differences between the measurements.

____

4. Describe in your own words the meaning of *meniscus*. Why is reading a meniscus at eye level important for measuring volume accurately? What else is important in order to measure the volume of liquids correctly?

____

____

____

____

DATASHEET

Finding Volume by Displacement

Teacher Notes

In this activity, students will learn how to find the volume of irregular objects (covers standard 8.8.b). Use large keys that can still fit into the cylinder. You may wish to use plastic graduated cylinders to avoid breakage. Water depth must be sufficient to submerge the key.

MATERIALS

For each group

- graduated cylinder, 100 mL
- key, metal
- water, colored

SAFETY CAUTION

Remind students to review all safety cautions and icons before beginning this activity.

(Quick Lab) **DATASHEET A**

Finding Volume by Displacement

SAFETY INFORMATION

PROCEDURE

1. Fill a **100 mL graduated cylinder** to about the 50 mL mark with **colored water**.
   - Read the volume at the meniscus. (Hint: The meniscus is the curve at the liquid's surface.)
   - Write down the volume here. ____________________________

2. Tip the graduated cylinder just a little.
   - Slide a **metal key** along the side of the cylinder into the water.
   - Don't let the water splash.

3. Carefully read the new level of water in the cylinder.
   - Write down your reading here. ____________________________

4. Calculate the volume of water displaced by the key. (Hint: Displaced water is water pushed out of the way.)
   - Show your work below.

   Volume of water with key (from step 3): ____________________________

   Volume of water without key (from step 1): ____________________________

   Volume of water displaced by key: ____________________________

5. Use your results to calculate the volume of the key. (Hint: The volume of water displaced by the key is equal to the key's volume.)
   - Be sure to use the correct units to describe the volume of this solid. (Hint: The volume of a solid should not be given in liters or milliliters, so remember to convert from mL into cm^3.)
   - Show your work below.

   Volume of water displaced by key
   (from step 4): ____________________________

   Convert mL to cubic units
   (Hint: $1\ mL = 1\ cm^3$): ____________________________

   Volume of key: ____________________________

Finding Volume by Displacement

SAFETY INFORMATION

PROCEDURE

1. Fill a **100 mL graduated cylinder** to about the 50 mL mark with **colored water.** Read the volume at the meniscus, and record your reading.

   ____

   ____

2. Tip the graduated cylinder, and slide a **metal key** into the cylinder.

3. Carefully read the new level of water in the cylinder, and record your results.

   ____

   ____

4. Calculate the volume of water displaced by the key. Show your work below.

5. Using your results, calculate the volume of the key. Show your work below.

Quick Lab)   **DATASHEET C**

Finding Volume by Displacement

SAFETY INFORMATION

PROCEDURE

1. Fill a **100 mL graduated cylinder** to about the 50 mL mark with **colored water.** Read the volume at the meniscus, and record your reading.

   ____

2. Tip the graduated cylinder and slide a **metal key** into the cylinder.

3. Carefully read the new level of water in the cylinder, and record your results.

   ____

4. Calculate the volume of water displaced by the key. Show your work below.

5. Using your results, calculate the volume of the key. What are the correct units to describe this volume? Explain why it is important to tip the cylinder before inserting the key.

   ____

   ____

   ____

   ____

   ____

   ____

Quick Lab **DATASHEET**

Finding the Density of Unknown Metals

Teacher Notes

In this activity, students will calculate the density of irregular objects (covers standards 8.8.a, 8.8.b, and 8.9.f). Objects should be large enough to displace an amount of water that students can easily measure. If you use pennies in this activity, separate pre-1982 pennies from post-1982 pennies. Pre-1982 pennies are solid copper and post-1982 pennies are copper-coated zinc. Using smaller graduated cylinders will give better results.

MATERIALS

For each group

• balance, triple-beam

• graduated cylinder

• metal objects, such as screws, washers, and ball bearings

• water

SAFETY CAUTION

Remind students to review all safety cautions and icons before beginning this lab activity. Dropping objects into a graduated cylinder can break it. Show students how to slide an object down the side of a tilted cylinder.

Quick Lab **DATASHEET A**

Finding the Density of Unknown Metals

In this activity, you will use measurements of mass and volume to calculate the density of several metal objects.

SAFETY INFORMATION

TRY IT!

1. Examine a **small metal object.**
   - Write down the properties that you observe in the data table.

2. Measure the mass of the object by using a **triple-beam balance.** (Hint: Make sure to set the balance to zero before you start.)
   - Write down the mass in the data table.

3. Fill a **graduated cylinder** about halfway with **water.**
   - Read the volume at the meniscus. (Hint: Make sure to look at eye level when reading the volume. Be sure to read the volume at the bottom of the meniscus.)
   - Write down your reading in the data table, in the column "Initial Volume."

4. Tip the cylinder.
   - Gently slide the metal object into the cylinder.
   - Read the new volume.
   - Write down your reading in the data table.

5. Calculate the volume of the object. Show your work below. (Hint: Make sure that your answer uses the correct units, $1 \text{ mL} = 1 \text{ cm}^3$)

   Volume of water with metal object
   (from step 4): _______________________

   Volume of water without metal object
   (from step 3): _______________________

   Volume of water displaced by metal object: _____________________________

   Volume of metal object (Hint: same as displaced volume):

   ____

   - Write down the volume in the data table.

Finding the Density of Unknown Metals *continued*

THINK ABOUT IT!

6. Repeat steps 1–5 for two other metal objects. Record your data in the data table.

Table Observations				
Object	Properties	Mass of Object	Initial Volume in Cylinder	Volume of Object
1				
2				
3				

7. Use your data to calculate the density of each metal object. (Hint: Convert mL to cm^3.)

Object 1: (mass) ____________ ÷ (volume) ____________ = ____________

Object 2: (mass) ____________ ÷ (volume) ____________ = ____________

Object 3: (mass) ____________ ÷ (volume) ____________ = ____________

8. Write the objects in order from least to greatest density.

(least density) __________ __________ __________ (greatest density)

9. Which of the objects is most likely made of lead? (Hint: Lead has a density of 11.35 g/cm^3.)

____

____

• Why did you choose the object you did for your answer?

____

____

Finding the Density of Unknown Metals

In this activity, you will use measurements of mass and volume to calculate the density of several metal objects.

SAFETY INFORMATION 

TRY IT!

1. Examine a **small metal object,** and record the properties that you observe.

____

2. Measure the mass of the object by using a **triple-beam balance.** Record the mass.

____

3. Fill a **graduated cylinder** about halfway with **water.** Read the volume at the meniscus, and record your reading.

____

4. Tip the cylinder, and gently slide the metal object into the cylinder. Read the new volume, and record your reading.

____

5. Calculate the volume of the object. Record your results. Show your work below.

6. Repeat steps 1–5 for each metal object.

Finding the Density of Unknown Metals *continued*

THINK ABOUT IT!

7. Use your data to calculate the density of each metal object. Show your work below.

8. List the objects in order of least to greatest density.

____

____

____

____

____

9. Is one or more of the objects most likely made of lead? Explain your answer. (Hint: Lead has a density of 11.35 g/cm^3.)

____

____

____

____

____

____

____

Quick Lab   **DATASHEET C**

Finding the Density of Unknown Metals

In this activity, you will use measurements of mass and volume to calculate the density of several metal objects.

SAFETY INFORMATION

TRY IT!

Create a table to record your data.

1. Examine a **small metal object**, and record the properties that you observe.

2. Measure the mass of the object by using a **triple-beam balance.** Record the mass.

3. Fill a **graduated cylinder** about halfway with **water.** Read the volume at the meniscus, and record your reading.

4. Tip the cylinder, and gently slide the metal object into the cylinder. Read the new volume, and record your reading.

5. Calculate the volume of the object. Record your results.

6. Repeat steps 1–5 for each metal object.

THINK ABOUT IT!

7. Use your data to calculate the density of each metal object.

8. Number the objects in your data table from least to greatest density.

9. Is one or more of the objects most likely made of lead? Explain your answer. (Hint: Lead has a density of 11.35 g/cm^3.)

____

____

____

Quick Lab

Physical or Chemical Change?

Teacher Notes

In this activity, students distinguish between chemical and physical changes (covers standard 8.5.a). Be sure students do not open the bag when using the magnet.

MATERIALS

For each group

- bag, plastic
- beaker
- effervescent tablet
- iron filings
- magnet, bar
- sand
- stick, wooden
- test tube
- water

SAFETY CAUTION

Remind students to review all safety cautions and icons before beginning this activity.

Quick Lab   **DATASHEET**

Physical or Chemical Change?

SAFETY INFORMATION

PROCEDURE

1. Watch as your teacher places a burning **wooden stick** into a **test tube.**
   - Write your observations on the lines below.

____

____

____

____

2. Place a mixture of **sand** and **iron filings** in a **plastic bag.**
   - Seal the bag.
   - Place a bar magnet on top of the bag.
   - Try to separate the iron from the sand.
   - Write your observations on the lines below.

____

____

____

____

3. Drop an **effervescent tablet** into a **beaker** of **water.**
   - Write your observations on the lines below.

____

____

____

____

____

Physical or Chemical Change? *continued*

4. For each step above, tell if the change is a physical change or a chemical change. Explain how you know.

a. Placing a burning wooden stick into a test tube results in a

_____________________ change because _____________________

____

b. Separating the iron from the sand with a bar magnet results in a

_____________________ change because _____________________

____

c. Dropping an effervescent tablet into a beaker of water results in a

_____________________ change because _____________________

____

Physical or Chemical Change?

SAFETY INFORMATION

PROCEDURE

1. Watch as your teacher places a burning **wooden stick** into a **test tube.** Record your observations.

____

____

2. Place a mixture of **sand** and **iron filings** in a **plastic bag,** and seal the bag. Place a **bar magnet** on top of the bag, and try to separate the iron from the sand.

3. Drop an **effervescent tablet** into a **beaker of water.** Record your observations.

____

____

4. For each step, identify which kind of change happens: a physical change or a chemical change. Explain your answers.

____

____

____

____

____

____

____

(Quick Lab)   **DATASHEET C**

Physical or Chemical Change?

SAFETY INFORMATION

PROCEDURE

1. Watch as your teacher places a burning **wooden stick** into a **test tube.**
Record your observations.

____

____

____

____

2. Place a mixture of **sand** and **iron filings** in a **plastic bag,** and seal the bag.
Place a **bar magnet** on top of the bag, and try to separate the iron from the
sand. Record your observations.

____

____

____

____

3. Drop an **effervescent tablet** into a **beaker** of **water.** Record your
observations.

____

____

____

____

4. For each step, identify what type of change occurs. Explain your answers.

____

____

____

____

____

Skills Practice Lab)

DATASHEET

Classifying Substances

Teacher Notes

In this lab, students observe chemical and physical properties of substances and use these properties to classify substances (covers standards 8.5.a and 8.7.c). To help students organize their data, have students number the egg-carton cups and put corresponding numbers on their data tables.

Anna Marcello
Audubon Middle School
Los Angeles, California

TIME REQUIRED

One 45-min class period

LAB RATINGS

Easy ← 1 2 3 4 → Hard

Teacher Prep–2
Student Set-Up–1
Concept Level–2
Clean Up–2

MATERIALS

The materials listed on this page are enough for groups of two or three students. Be sure that egg cartons do not have cracks in the cups. The iodine solution should be a water solution, not tincture of iodine. Dilute the 2% iodine solution used for biological staining about 1:1 with water. A small test tube taped to a bottle makes a great holder for a dropper or pipette and helps prevent cross-contamination. A large plastic drinking straw cut in half at a shallow angle works well as a spatula.

SAFETY CAUTION

Remind students to review all safety cautions and icons before beginning this lab activity. When iodine is being used, be certain that a functioning eyewash is available. Caution students that iodine can stain skin and clothes. Students should wash their face and hands when finished. Clean up any spills immediately to avoid slips and falls. Remind students that vinegar is an acid, and they should use caution when handling it.

Classifying Substances

You have learned how to describe matter based on its physical and chemical properties. You have also learned signs that can help you determine whether a change in matter is a physical change or a chemical change. In this lab, you will use what you have learned to describe four substances and their properties based on the changes that the substances undergo.

OBJECTIVES

Describe the physical properties of four substances.

Identify physical and chemical changes.

Classify four substances by their chemical properties.

MATERIALS

- baking powder
- baking soda
- carton, egg, plastic-foam
- cornstarch
- eyedroppers (3)
- iodine solution

- spatulas, metal (4)
- stirring rod
- sugar
- vinegar
- water

SAFETY INFORMATION  

PROCEDURE

1. Use Table 1 and Table 2 to record your data.

2. Start with the baking powder.
   - Use a spatula to place a small amount of baking powder into three cups of your egg carton.
   - Use just enough baking powder to cover the bottom of each cup.
   - Look at the baking powder.
     - What color is it?
     - What is its texture? Does it look smooth, or rough?
   - Write your answers in Table 1 in the column labeled "Unmixed."

Classifying Substances *continued*

3. Work with the baking powder in the first cup of the egg carton.
   - Use an eyedropper to add 60 drops (about 3 mL) of water to the baking powder in the first cup.
   - Stir with the stirring rod.
   - Look at the mixture.
   - What is the color. Does a reaction take place?
   - Write what you see in Table 1 in the column labeled "Mixed with water."
   - Clean your stirring rod.

4. Now work with the baking powder in the second cup of the egg carton.
   - Use a clean dropper.
   - Add 20 drops of vinegar to the second cup of baking powder.
   - Stir.
   - Write what you see in Table 1 in the column labeled "Mixed with vinegar."
   - Clean your stirring rod.

5. Next, work with the baking powder in the third cup of the egg carton.
   - Use a clean dropper.
   - Add 5 drops of iodine solution to the third cup of baking powder.
   - **Caution**: Be careful when using iodine. Iodine will stain your skin and clothes.
   - Stir.
   - Write what you see in Table 1 in the column labeled "Mixed with iodine solution."
   - Clean your stirring rod.

Table 1 Observations

Substance	Unmixed	Mixed with water	Mixed with vinegar	Mixed with iodine solution
Baking powder				
Baking soda				
Cornstarch				
Sugar				

Classifying Substances *continued*

6. Repeat steps 2–5 for each of the other substances (baking soda, cornstarch, and sugar).
  - Use a clean spatula for each substance.
  - Check off your steps in the following chart as you work.

Steps	Baking Soda	Cornstarch	Sugar
• Put substance in three egg carton cups.			
• Describe in Table 1: "Unmixed."			
First cup			
• Add 60 drops of water.			
• Stir.			
• Describe in Table 1: "Mixed with water."			
• Clean your stirring rod.			
Second cup			
• Use a clean dropper.			
• Add 20 drops of vinegar.			
• Stir.			
• Describe in Table 1: "Mixed with vinegar."			
• Clean your stirring rod.			
Third cup			
• Use a clean dropper.			
• Add 5 drops of iodine solution.			
• Stir.			
• Describe in Table 1: "Mixed with iodine solution."			
• Clean your stirring rod.			

Classifying Substances *continued*

ANALYZE THE RESULTS

7. Analyzing Data In Table 2, write down the following for each substance:
- The type of change you saw—physical or chemical
- The property that the change demonstrates—reactivity or solubility

Table 2 Changes and Properties

Substance	Mixed with water		Mixed with vinegar		Mixed with iodine solution	
	Change	Property	Change	Property	Change	Property
Baking powder						
Baking soda						
Cornstarch						
Sugar						

DRAW CONCLUSIONS

8. Making Predictions Suppose that you have a mixture of two substances.
- The mixture turns black when it is mixed with iodine and bubbles when it is mixed with water and vinegar.
- What two substances are in the mixture? (Hint: Use two of the substances from this experiment.)

____

____

____

____

BIG IDEA QUESTION

9. Drawing Conclusions Reactivity is a chemical property. Tell which liquid each of the four substances reacted with, if any.

____

____

____

____

Classifying Substances

You have learned how to describe matter based on its physical and chemical properties. You have also learned signs that can help you determine whether a change in matter is a physical change or a chemical change. In this lab, you will use what you have learned to describe four substances and their properties based on the changes that the substances undergo.

OBJECTIVES

Describe the physical properties of four substances.

Identify physical and chemical changes.

Classify four substances by their chemical properties.

MATERIALS

- baking powder
- baking soda
- carton, egg, plastic-foam
- cornstarch
- eyedroppers (3)
- iodine solution

- spatulas, metal (4)
- stirring rod
- sugar
- vinegar
- water

SAFETY INFORMATION 

PROCEDURE

1. Copy Table 1 and Table 2 shown on the next page. Be sure to leave plenty of room in each box to write down your observations.

2. Use a spatula to place a small amount of baking powder into three cups of your egg carton. Use just enough baking powder to cover the bottom of each cup. Record your observations about the baking powder's appearance, such as color and texture, in Table 1 in the column labeled "Unmixed."

3. Use an eyedropper to add 60 drops (about 3 mL) of water to the baking powder in the first cup. Stir with the stirring rod. Record your observations in Table 1 in the column labeled "Mixed with water." Clean your stirring rod.

4. Use a clean dropper to add 20 drops of vinegar to the second cup of baking powder. Stir. Record your observations in Table 1 in the column labeled "Mixed with vinegar." Clean your stirring rod.

5. Use a clean dropper to add 5 drops of iodine solution to the third cup of baking powder. Stir. Record your observations in Table 1 in the column labeled "Mixed with iodine solution." Clean your stirring rod. **Caution:** Be careful when using iodine. Iodine will stain your skin and clothes.

Classifying Substances *continued*

6. Repeat steps 2–5 for each of the other substances (baking soda, cornstarch, and sugar). Use a clean spatula for each substance.

Table 1 Observations

Substance	Unmixed	Mixed with water	Mixed with vinegar	Mixed with iodine solution
Baking powder				
Baking soda				
Cornstarch				
Sugar				

Table 2 Changes and Properties

Substance	Mixed with water		Mixed with vinegar		Mixed with iodine solution	
	Change	Property	Change	Property	Change	Property
Baking powder						
Baking soda						
Cornstarch						
Sugar						

ANALYZE THE RESULTS

7. Analyzing Data In Table 2, write the type of change that you observed for each substance (physical or chemical). State the property that the change demonstrates.

Classifying Substances *continued*

DRAW CONCLUSIONS

8. Making Predictions Suppose that you have a mixture of two substances. The mixture turns black when it is mixed with iodine and bubbles when it is mixed with water and vinegar. What two substances are in the mixture?

____

____

____

____

____

____

____

BIG IDEA QUESTION

9. Drawing Conclusions How can you describe each of the four substances in terms of the chemical property of reactivity?

____

____

____

____

____

____

____

____

____

____

Classifying Substances

You have learned how to describe matter based on its physical and chemical properties. You have also learned signs that can help you determine whether a change in matter is a physical change or a chemical change. In this lab, you will use what you have learned to describe four substances and their properties based on the changes that the substances undergo.

OBJECTIVES

Describe the physical properties of four substances.

Identify physical and chemical changes.

Classify four substances by their chemical properties.

MATERIALS

- baking powder
- baking soda
- carton, egg, plastic-foam
- cornstarch
- eyedroppers (3)
- iodine solution
- spatulas, metal (4)
- stirring rod
- sugar
- vinegar
- water

SAFETY INFORMATION

PROCEDURE

1. Read through the entire lab. Then, create a table to record your observations. Be sure to leave plenty of room for writing your observations.

2. Use a spatula to place a small amount of baking powder into three cups of your egg carton. Use just enough baking powder to cover the bottom of each cup. Record your observations about the baking powder's appearance, such as color and texture, in your table.

3. Use an eyedropper to add 60 drops (about 3 mL) of water to the baking powder in the first cup. Stir with the stirring rod. Record your observations in your table. Clean your stirring rod.

4. Use a clean dropper to add 20 drops of vinegar to the second cup of baking powder. Stir. Record your observations in your table. Clean your stirring rod.

5. Use a clean dropper to add 5 drops of iodine solution to the third cup of baking powder. Stir. Record your observations in your table. Clean your stirring rod.
Caution: Be careful when using iodine. Iodine will stain your skin and clothes.

6. Repeat steps 2–5 for each of the other substances (baking soda, cornstarch, and sugar). Use a clean spatula for each substance.

Classifying Substances *continued*

ANALYZE THE RESULTS

7. Analyzing Data Create a table to record the type of change—physical or chemical—that you observed for each substance. State the property that the change demonstrates.

DRAW CONCLUSIONS

8. Making Predictions Suppose that you have a mixture of two substances. The mixture turns black when it is mixed with iodine and bubbles when it is mixed with water and vinegar. What two substances are in the mixture?

____

____

____

BIG IDEA QUESTION

9. Drawing Conclusions How can you describe each of the four substances in terms of the chemical property of reactivity?

____

____

____

____

 DATASHEET

Using a Three-Variable Equation

Teacher Notes

In this activity, students will learn how to rearrange a three variable equation and will use a three-variable equation to solve problems (covers standard 8.9.f).

Using a Three-Variable Equation

INVESTIGATION AND EXPERIMENTATION

8.9.f Apply simple mathematical relationships to determine a missing quantity in a mathematic expression, given the two remaining terms (including speed = distance/time, density = mass/ volume, force = pressure × area, volume = area × height).

TUTORIAL
Procedure

A three-variable equation can be rearranged and used to solve for any of the variables in the equation. The equation for density is an example of a three-variable equation.

1. The equation for density is arranged to calculate density.

$$D = \frac{m}{V}$$

2. To rearrange the equation to solve for mass, multiply both sides of the equation by volume.

$$m = D \times V$$

3. To rearrange the equation to solve for volume, divide both sides of the equation for mass by density.

$$V = \frac{m}{D}$$

Analysis

4. Using Equations To determine which form of the equation you should use to solve a problem, read the question carefully. The question will ask you to find one of the variables. Choose the form of the equation that is used to solve for that variable.

YOU TRY IT!
Procedure

Use the equation for density to answer the following questions. Show all of your work, including the work that you do to rearrange the equation, if necessary.

Using a Three-Variable Equation *continued*

Analysis

1. Using Equations If a 16.9 cm^3 cube of ice has a mass of 15.5 g, what is the density of the ice?

2. Using Equations A piece of metal has a density of 11.3 g/cm^3 and a volume of 6.7 cm^3. What is the mass of this piece of metal?

3. Using Equations Tin has a density of 7.31 g/cm^3. What is the volume of a piece of tin whose mass is 18.3 g?

4. Using Equations Suppose that the 12-sided object shown in the image at right has a mass of 5.4 g. What is the density of the object?

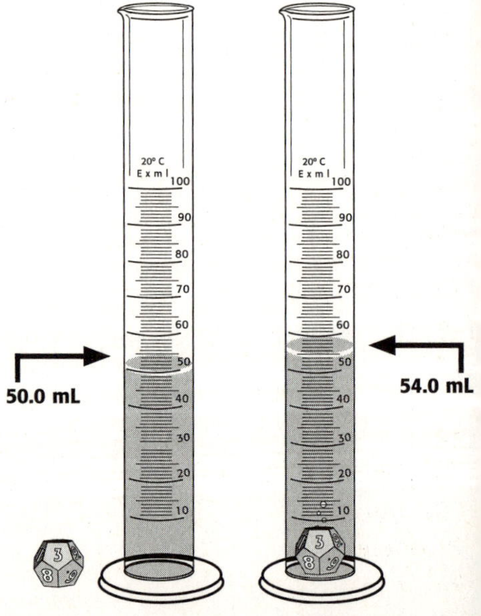

Answer Key

Directed Reading A

SECTION: WHAT IS MATTER?

1. C
2. B
3. C
4. A
5. B
6. D
7. meniscus
8. cubic
9. volume
10. irregular solid
11. milliliter
12. cubic centimeters
13. B
14. C
15. B
16. D
17. C
18. D
19. A
20. mass
21. kilogram
22. newton

SECTION: PHYSICAL PROPERTIES

1. A
2. C
3. C
4. B
5. A
6. B
7. C
8. A
9. A
10. B
11. D
12. B
13. D
14. A
15. B
16. D
17. B
18. C
19. physical change
20. state
21. identity
22. B
23. C
24. C
25. B

SECTION: CHEMICAL PROPERTIES

1. chemical property
2. reactivity
3. flammability
4. nonflammability
5. B
6. C
7. A
8. B
9. D
10. C
11. property
12. change
13. B
14. C
15. A
16. C
17. D
18. B
19. B
20. A

Directed Reading B

SECTION: WHAT IS MATTER?

1. They are all made of matter.
2. Matter is anything that has mass and takes up space.
3. B
4. D
5. Volume is the amount of space taken up by an object (or the size of a region of space).
6. volume
7. meniscus
8. cubic
9. length, width, and height
10. Answers may vary. Sample answer: The volume could be measured by submerging the object in a graduated cylinder with water. The volume of water displaced is the volume of the object.
11. 8 cm^3
12. D
13. C
14. C

15. D

16. The only way to change the mass of an object is to change the amount of matter the object contains.

17. mass

18. weight

19. mass

20. weight

21. weight

SECTION: PHYSICAL PROPERTIES

1. B

2. C

3. D

4. A

5. F

6. C

7. B

8. E

9. G

10. physical property

11. density

12. $D = \dfrac{m}{V}$

13. density; volume; mass

14. volume

15. The density will not change because the density of a given substance remains the same no matter how much of it you have.

16. Answers may vary. Sample answer: because a substance's density is always the same at a given temperature and pressure and because each substance has a unique density

17. because 1 kg of lead would take up much less space than 1 kg of feathers

18. The object will sink.

19. Answers may vary. Sample answer: If you know the density of the substance, you can compare it with the density of water. If the density of the object is less than the density of water, it will float.

20. densities

21. The densest layer will settle on the bottom.

22. The layer with the least density will be found on top.

23. physical change

24. physical changes

25. PC

26. X

27. PC

28. PC

29. PC

30. PC

31. identity

32. Answers may vary. Sample answer: When matter undergoes a physical change, one or more physical properties are changed. For example, if a lump of copper is drawn out into a thin wire, its shape is changed, but not its identity.

SECTION: CHEMICAL PROPERTIES

1. C

2. A

3. B

4. D

5. B

6. Answers may vary. Sample answer: The burning changes wood to smoke and ashes.

7. chemical

8. characteristic

9. B

10. C

11. Answers may vary. Sample answer: Baking a cake involves chemical changes because the cake has completely different properties than its original ingredients.

12. Answers may vary. Sample answer: The creation of new substances with new properties shows that a chemical change is taking place. Other signs include fizzing or foaming, a change in color or odor, and the production of heat, sounds, or light.

13. liberates *(or releases)*

14. absorbs

15. chemical

16. Answers may vary. Sample answer: Some chemical changes can be reversed with more chemical changes. For example, the water formed in a space shuttle's rockets can later be split back into hydrogen and oxygen using an electric current.

17. B

18. A

19. C

20. CC

21. PC

22. CC

23. PC
24. CC
25. PC
26. PC

Vocabulary and Section Summary A

SECTION: WHAT IS MATTER?

1. matter: anything that has mass and takes up space
2. volume: a measure of the size of a body or region in three-dimensional space
3. meniscus: the curve at a liquid's surface by which one measures the volume of the liquid
4. mass: a measure of the amount of matter in an object
5. weight: a measure of the gravitational force exerted on an object; its value can change with the location of the object in the universe

SECTION: PHYSICAL PROPERTIES

1. physical property: a characteristic of a substance that does not involve a chemical change, such as density, color, or hardness
2. density: the ratio of the mass of a substance to the volume of the substance
3. physical change: a change of matter from one form to another without a change in chemical properties

SECTION: CHEMICAL PROPERTIES

1. chemical property: a property of matter that describes a substance's ability to participate in chemical reactions
2. chemical change: a change that occurs when one or more substances change into entirely new substances with different properties

Vocabulary and Section Summary B

SECTION: WHAT IS MATTER?

Across

2. meniscus
3. weight
4. volume

Down

1. mass
2. matter

SECTION: PHYSICAL PROPERTIES

1. physical property
2. density
3. physical change
4. volume
5. mass

W	I	M	M	Q	W	T	I	D	I	R	X	D	P	V	Y	Y	C	H	H
J	T	A	S	Y	T	V	F	R	B	Q	T	L	E	T	Y	R	Y	D	H
K	S	L	O	K	L	Q	P	Z	W	D	X	S	R	N	Z	G	O	X	B
S	B	C	N	R	C	U	I	W	C	F	G	E	A	D	S	V	H	N	A
F	H	C	S	G	H	C	F	P	U	C	P	T	M	R	X	I	A	U	A
R	R	D	G	G	P	V	S	M	K	O	R	O	Z	X	I	P	T	O	W
Z	H	M	Z	D	S	Y	Q	C	R	C	E	A	B	E	C	E	I	Y	N
B	S	Q	C	B	M	Y	J	P	A	R	K	E	I	Y	N	I	P	B	P
E	Z	N	O	K	G	V	L	Q	L	M	L	Y	M	Z	N	N	L	B	O
L	Q	U	O	L	D	A	E	B	G	Y	F	W	D	R	Z	R	L	Q	V
G	P	S	B	M	C	L	X	S	B	L	Z	Q	E	P	G	X	O	I	U
G	R	Q	X	I	A	K	Q	O	E	Q	F	N	Q	I	X	H	Y	W	H
P	Y	K	S	G	A	W	U	X	B	S	Z	K	S	K	B	V	A	Q	U
A	D	Y	Q	X	F	X	I	C	K	M	I	B	W	I	I	O	O	X	W
D	H	P	H	Y	S	I	C	A	L	C	H	A	N	G	E	L	D	M	K
P	X	B	H	F	T	S	X	B	S	Z	D	P	P	B	X	U	N	M	O
A	O	H	P	P	W	N	O	G	U	J	N	A	O	B	Q	M	F	P	M
E	P	U	X	A	Z	G	D	P	R	D	V	T	S	T	B	E	B	W	A
E	K	G	B	U	P	W	D	G	M	P	S	C	H	O	A	X	Y	J	L
M	B	X	P	A	S	H	K	U	Z	C	N	Y	E	U	Y	P	V	J	X

SECTION: CHEMICAL PROPERTIES

1. chemical property
2. reactivity
3. chemical change
4. characteristic properties
5. flammability
6. composition

Reinforcement

A MATTER OF DENSITY

1. Green liquid: 0.75 kg/L; Blue liquid: 0.9 kg/L; Red liquid: 1.2 kg/L; Black liquid: 0.8 kg/L
2. First (bottom): red; Second: blue; Third: black; Fourth (top): green
3. B
4. The layers of the diagram should be shaded/labeled in the following order from the top: green, black, blue, red.
5. Accept all reasonable answers. Sample answer: I could open the spigot at the bottom of the tank and let the red liquid out.

Critical Thinking

1. The cube's volume increased, and its mass remained the same.

2. Answers may vary. Sample answer: Because the cube's volume increased and its mass remained the same, the larger cube has less mass per unit volume. As a result, the cube's density decreased.
3. Answers may vary. Sample answer: The cube changed in volume, density, mass, number, and composition.
4. Answers may vary. Sample answer: The changes in volume, density and number are physical changes because the identity of the cube remained the same. The cube's transformation into a green paste is a chemical change because a new substance was formed.
5. Answers may vary. Sample answer: An object's mass or volume does not usually change on Earth spontaneously without a recognizable cause. An object's density typically remains the same at a given temperature and pressure.
6. Answers may vary. Sample answer: An explosion caused by electricity could happen on Earth. Applying an electric current causes some chemical changes such as splitting water into the gases that make it up.

SciLinks Activity

1. Answers may vary. Sample answer: Materials science covers a broad range of activities and touches on many scientific fields such as chemistry, biology, and physics. Materials scientists are sometimes engineers or metallurgists. Their job is to manipulate or change materials based on our understanding of how the materials are put together.
2. Answers may vary. Sample answer: *Polymers* are long-repeating chains of smaller molecules linked together. Rubber bands, paints, and plastics are made from polymers. *Composites* are combinations of different kinds of materials. A bicycle might be made from an ultralight carbon-fiber composite. *Semiconductors* conduct electricity. They are often used in electronic computer chips. *Metals* were among the first substances to be engineered. They can be drawn into wires and molded, and they conduct electricity very well. *Biomaterials* are found in every part of your body from your bones to your fingernails and hair. Their properties can protect you from heat or cold or help you to cut and grind food.
3. Answers may vary. Sample answer: Water has a high surface tension. It has the ability to form drops. Water expands and gets less dense when it freezes and will actually float on its liquid form. Many things can be dissolved in liquid water. Because of its hydrogen bonds, water boils at an extremely high temperature and freezes at a much higher temperature than might be expected for the size of its particles. Also, because of its bonding, water has a very high specific heat, which means that it can absorb a lot of heat before it begins to get hot.

Section Review

SECTION: WHAT IS MATTER?

1. Sample answer: You can measure the volume of a liquid by pouring it into a graduated cylinder and reading the scale at the bottom of the meniscus.
2. Sample answer: Matter is anything that has mass and takes up space. Mass is a measurement of how much matter is in an object. Weight is a measurement of how much gravitational force is exerted on an object's mass.
3. Volumes of solids can be expressed in cubic meters or cubic centimeters.
4. $225 \text{ mL} - 85 \text{ mL} = 140 \text{ mL}$

5. Not all objects that have large masses have large weights because the weight of an object can change depending on where it is located in the universe. Mass remains the same everywhere in the universe. So, a massive object in space may not have a large weight.

6. volume = area × height
   B1: 36 m^2 × 4 m = 144 m^3
   B2: 16 m^2 × 9 m = 144 m^3

7. area = volume ÷ height =
   144 m^3 ÷ 12 m = 12 m^2

8. The elephant's weight on the moon would be less than its weight on Earth because the gravitational force of the moon is less than the gravitational force of Earth.

SECTION: PHYSICAL PROPERTIES

1. Sample answer: A physical property can be observed without changing the identity of the matter. When matter undergoes a physical change, its shape or form changes, but its identity remains the same.

2. A golf ball feels heavier than a table-tennis ball because a golf ball is denser; that is, it has more mass in a similar volume.

3. A physical change changes the shape or form of the matter without changing its identity. It is still the same matter as before the change and has some of the same properties.

4. 26 mL = 26 cm^3; D = m ÷ V =
   273 g ÷ 26 cm^3 = 10.5 g/cm^3;
   The substance is silver.

5. V = m ÷ D = 408 g ÷ 11.35 g/cm^3 =
   35.95 cm^3 = 35.95 mL

6. Use the mass and volume of the coin to calculate the coin's density. Compare this density with the known density of silver.

7. The substances are all liquids.

8. D = m ÷ V = 350 g ÷ 95 cm^3 = 3.68 g/cm^3; This object would not float in water because its density is greater than the density of water.

9. Sample answer: I would find the mass of a clean, dry graduated cylinder, pour some of the unknown liquid into the graduated cylinder, and note the volume. Then, I would find the mass of the graduated cylinder with the liquid in it. Next, I would subtract the mass of the empty graduated cylinder from the mass of the graduated cylinder with the liquid in it to find the mass of the liquid. Finally, I would divide the mass of the liquid by the volume of the liquid to calculate the density.

SECTION: CHEMICAL PROPERTIES

1. Sample answer: A chemical property is a property that describes matter based on its ability to change into new matter. A chemical change happens when one or more substances are changed into new substances.

2. Rusting is a chemical change because the iron is reacting with oxygen to make a new substance, rust, that has different chemical properties.

3. One example of a chemical property is the reactivity of iron with oxygen to form rust. Reactivity tells you whether a substance will react with another substance and form new products with new properties. Another example is the flammability of firewood to form ash, carbon dioxide, water vapor, heat, and light. Flammability tells you whether a substance will burn.

4. A chemical change occurred. The copper in the Statue of Liberty combined with substances in the air and formed new copper compounds that have different properties.

5. Heat can be liberated or absorbed during a chemical change.

6. Sample answer: The composition of a substance is altered during a chemical change. The composition of a substance stays the same during a physical change.

7. Melting is a physical change because the composition of a substance doesn't change during melting.

8. temperature change = (3 min × 2°C/min) + 2°C = 8°C; final temperature = 25°C + 8°C = 33°C
9. Sample answer: Wax burning is a chemical change. I know that this is a chemical change because heat is liberated. The wick burning is another chemical change. The wick changes into new substances that have different properties.

Chapter Review

1. Liberate means "to give off."
2. Sample answer: A physical property of a substance is a characteristic such as color, shape, or density. Reactivity is a chemical property of matter. Sugar dissolving in water is an example of a physical change. When a piece of iron rusts, it is undergoing a chemical change.
3. Sample answer: Mass is the amount of matter in an object and does not change with location. Weight is a measure of the gravitational force and will change depending on an object's location in the universe.
4. Sample answer: Volume is the amount of space occupied by an object, and density is the amount of mass in a given volume.
5. B
6. A
7. D
8. C
9. B
10. Sample answer: Two characteristic properties are density and reactivity.
11. Sample answer: The volume of a liquid can be measured by pouring the liquid into a graduated cylinder and reading the scale at the bottom of the meniscus. The volume of a rectangular solid can be determined by multiplying the object's length, width, and height. The volume of an irregular solid can be measured using the process of water displacement.
12. Dish detergent is more dense than water, but less dense than maple syrup.
13. Corn oil would be on the top, shampoo would be in the middle, and antifreeze would be on the bottom.

14. Sample answer: Three properties of the can are a crushed shape, a somewhat shiny finish, and a red color.
15. The can underwent a physical change because the composition of the can did not change.
16. The density of the metal before and after the change is the same because density is a characteristic property of matter.
17. No, chemical properties cannot be determined simply by looking at a substance. Chemical properties can only be observed when a chemical change might occur.

Section Quizzes

SECTION: WHAT IS MATTER?
1. B
2. E
3. D
4. C
5. A
6. A
7. B
8. A
9. C

SECTION: PHYSICAL PROPERTIES
1. E
2. D
3. F
4. A
5. C
6. B
7. B
8. D
9. C
10. A

SECTION: CHEMICAL PROPERTIES
1. D
2. C
3. B
4. E
5. A
6. A
7. D
8. C
9. D
10. B

Chapter Test A

1. D
2. D
3. C
4. A
5. B
6. C
7. C
8. A
9. D
10. C
11. B
12. A
13. B
14. A
15. D
16. C
17. volume
18. property
19. change
20. mass
21. kilogram
22. weight
23. meniscus
24. milliliters

Chapter Test B

1. D
2. A
3. C
4. B
5. C
6. B
7. D
8. C
9. B
10. A
11. D
12. A
13. C
14. F
15. J
16. E
17. B
18. H
19. C
20. D
21. A
22. A
23. D
24. A
25. B

Chapter Test C

1. meniscus
2. density
3. physical change
4. chemical
5. weight
6. chemical change
7. C
8. D
9. D
10. B
11. C
12. Answers may vary. Sample answer: Characteristic properties such as density, reactivity, or solubility do not depend on the amount of the substance. Volume and mass cannot be characteristic properties because they are defined by measurements that depend on the size or amount of the substance in a sample.
13. Answers may vary. Sample answer: If the two substances are liquids, you could observe how they separate into layers as a result of differences in density. If the two substances are solids, you could observe whether they float or sink in water. Density is a useful property for identifying substances because most substances have unique densities that are constant at a given temperature and pressure.
14. Answers may vary. Sample answer: The astronauts should be able to carry more mass on the moon because the gravitational force acting on the objects is less there. Thus, the equipment will weigh less than it does on Earth.
15. Answers may vary. Sample answer: Most chemical changes involve either the liberation or absorption of heat. When heat is liberated, there is an increase in the surrounding temperature. When heat is absorbed, the surrounding temperature decreases.
16. Answers may vary. Sample answer: No; even though it would be nearly impossible to put the wood chips back together, this is still a physical change. The chips, log, and stump keep their identity as wood. Only their size and shape change.

17. Answers may vary. Sample answer: Mass is a measure of the amount of matter, while weight is a measure of gravitational force. Mass is constant no matter where the object is located, whereas weight depends on where it is found in space. Mass is measured with a balance and expressed in grams, whereas weight is measured with a spring scale and measured in newtons. People tend to confuse the two concepts because both the mass and the weight of a given object typically remain constant on Earth, and weight does give you a fairly good estimate of the mass of an object.

18. The purple liquid will form the bottom layer, followed by the water and the red liquid. The yellow liquid will be on top. The greater the density, the lower the layer. Water has a density of 1.00 g/cm^3, which means it will float on the liquid with a density of 1.62 g/cm^3. The wood chip will sink through the yellow liquid and float on top of the red liquid.

19. a. physical changes, **b.** chemical changes, **c.** identity, **d.** dissolving, **e.** substances, **f.** burning

Performance-Based Assessment

1. Sample answer: The luster and appearance of the pennies resemble the luster and appearance of the copper sample. The luster and appearance of the nickels, dimes, and quarters resemble the luster and appearance of the zinc sample.

2. Sample answer: I think the penny is made of copper because it is the same color as the copper sample. I think the nickel, dime, and quarter are mostly made of zinc because they are almost the same color as the zinc sample.

3-8. Values in student charts may vary, except for the densities of copper and zinc, which should be close to 8.96 and 7.13, respectively.

10. The penny is made mostly of zinc (unless the penny was made before 1982; then it is mostly copper). The nickel, dime, and quarter are made mostly of copper. The densities of the coins are close to the densities of the pure metals that primarily compose the coins.

Explore Activity

DATASHEET A

5. The units are grams. No, the amount of matter is not the same. The golf ball has more matter than the table-tennis ball.

6. No, if two objects are similar in size, it does not mean that they have the same amount of matter. The masses of two objects that have similar volumes depend on the materials out of which the objects are made. The golf ball and the table-tennis ball have similar sizes, but the golf ball has more mass and therefore more matter than the table-tennis ball has.

DATASHEET B

5. The units are grams. The golf ball has more matter than the table-tennis ball.

6. No. The masses of two objects that have similar volumes depend on the materials out of which the objects are made. The golf ball and the table-tennis ball have similar sizes, but the golf ball has more mass and therefore more matter than the table-tennis ball has.

DATASHEET C

5. The units are grams. The golf ball has more matter than the table-tennis ball.

6. No, if two objects are similar in size, it does not mean that they have the same amount of matter. The masses of two objects that have similar volumes depend on the materials out of which the objects are made. The golf ball and the table-tennis ball have similar sizes, but the golf ball has more mass and therefore more matter than the table-tennis ball has. Sample answers: a brick and a box of crackers; an orange and a tennis ball; a furnace and an empty refrigerator box; a photo album full of pictures and a box of cereal.

Quick Lab: Finding Accurate Volumes

DATASHEET A

3. Sample answer: When I read the graduated cylinder from below eye level, the volume appears to be less than 50 mL. When I read the graduated cylinder from above eye level, the volume appears to be more than 50 mL.
4. Sample answer: Reading the meniscus at an angle gives an incorrect reading.

DATASHEET B

3. Sample answer: When I read the graduated cylinder from below eye level, the volume appears to be less than 50 mL. When I read the graduated cylinder from above eye level, the volume appears to be more than 50 mL.
4. Sample answer: Reading the meniscus at an angle gives an incorrect reading.

DATASHEET C

3. Sample answer: When I read the graduated cylinder from below eye level, the volume appears to be less than 50 mL. When I read the graduated cylinder from above eye level, the volume appears to be more than 50 mL.
4. Sample answer: A meniscus is the curve at the surface of a liquid and is used to measure the liquid's volume. Reading the meniscus at eye level is important for obtaining an accurate measurement of liquid volume. If the meniscus is read at an angle or from different viewing positions, you will get an incorrect reading. When measuring a liquid's volume, it is also important to use a graduated cylinder instead of a measuring cup and to read the scale at the lowest point of the meniscus at eye level.

Quick Lab: Finding Volume by Displacement

DATASHEET A

4. Sample answer: 55 mL − 50 mL = 5 mL of water displaced

5. Sample answer: The volume of the key is 5 cm^3.

DATASHEET B

4. Sample answer: 55 mL − 50 mL = 5 mL of water displaced
5. Sample answer: The volume of the key is 5 cm^3.

DATASHEET C

4. Sample answer: 55 mL − 50 mL = 5 mL of water displaced
5. Sample answer: The volume of the key is 5 cm^3. The correct units to describe this volume are cubic centimeters. If the water is allowed to splash, some of it might spill out of the cylinder, resulting in an incorrect measurement of the volume of the key.

Quick Lab: Finding the Density of Unknown Metals

DATASHEET A

5. Answers may vary. Students should use the correct units for giving the volumes of liquids and the volumes of solids.
6. Answers may vary. Students should use the correct units for giving the volumes of liquids and the volumes of solids.
7. Students should correctly use the equation for density to find the density of each object. Be sure students convert mL to cm^3.
8. Answers may vary.
9. Answers may vary

DATASHEET B

7. Students should correctly use the equation for density to find the density of each object. Be sure students convert mL to cm^3.
8. Answers may vary.
9. Answers may vary.

DATASHEET C

7. Students should correctly use the equation for density to find the density of each object. Be sure students convert mL to cm^3.
8. Answers may vary.
9. Answers may vary.

Quick Lab: Physical or Chemical Change?

DATASHEET A

4. a. Placing a burning wooden stick into a test tube results in a chemical change because new substances with different properties are formed.
b. Separating the iron from the sand with a bar magnet results in a physical change because the sand and iron do not become new substances.
c. Dropping an effervescent tablet into a beaker of water results in a chemical change because a gas is produced.

DATASHEET B

4. Burning the stick and putting the tablet in water are chemical changes because new substances form. Separating iron filings from sand is a physical change because no new substances form.

DATASHEET C

4. Burning the stick and putting the tablet in water are chemical changes because new substances form. Separating iron filings from sand is a physical change because no new substances form.

Chapter Lab

DATASHEET A

7. See the table below.

Substance	Mixed with water		Mixed with vinegar		Mixed with iodine solution	
	Change	Property	Change	Property	Change	Property
Baking powder	chemical	reactivity with water	chemical	reactivity with acid	physical	solubility
Baking soda	physical	solubility	chemical	reactivity with acid	physical	solubility
Cornstarch	physical	solubility	physical	solubility	chemical	reactivity with iodine
Sugar	physical	solubility	physical	solubility	physical	solubility

8. Baking powder and cornstarch are in the mixture. Cornstarch is the only substance that reacts with iodine, and baking powder is the only substance that reacts with water.

9. Baking powder reacts when placed in water or vinegar. Cornstarch reacts with iodine to form a blue-black substance. Sugar does not react with water, vinegar, or iodine.

DATASHEET B

7. See the table below.

Substance	Mixed with water		Mixed with vinegar		Mixed with iodine solution	
	Change	Property	Change	Property	Change	Property
Baking powder	chemical	reactivity with water	chemical	reactivity with acid	physical	solubility
Baking soda	physical	solubility	chemical	reactivity with acid	physical	solubility
Cornstarch	physical	solubility	physical	solubility	chemical	reactivity with iodine
Sugar	physical	solubility	physical	solubility	physical	solubility

8. Baking powder and cornstarch are in the mixture. Cornstarch is the only substance that reacts with iodine, and baking powder is the only substance that reacts with water.

9. Baking powder reacts when placed in water or vinegar. Baking soda reacts with vinegar. Cornstarch reacts with iodine to form a blue black substance. Sugar does not react with water, vinegar, or iodine.

DATASHEET C

1.

Table 1 Observations				
Substance	Unmixed	Mixed w/ water	Mixed w/ vinegar	Mixed w/ iodine solution
Baking powder				
Baking soda				
Cornstarch				
Sugar				

7. See the table below.

Substance	Mixed with water		Mixed with vinegar		Mixed with iodine solution	
	Change	Property	Change	Property	Change	Property
Baking powder	chemical	reactivity with water	chemical	reactivity with acid	physical	solubility
Baking soda	physical	solubility	chemical	reactivity with acid	physical	solubility
Cornstarch	physical	solubility	physical	solubility	chemical	reactivity with iodine
Sugar	physical	solubility	physical	solubility	physical	solubility

8. Baking powder and cornstarch are in the mixture. Cornstarch is the only substance that reacts with iodine, and baking powder is the only substance that reacts with water.

9. Baking powder reacts when placed in water or vinegar. Cornstarch reacts with iodine to form a blue-black substance. Sugar does not react with water, vinegar, or iodine.

Science Skills Activity

DATASHEET

1. $D = 15.5 \text{ g} \div 16.9 \text{ cm}^3 = 0.917 \text{ g/cm}^3$
2. $m = 11.3 \text{ g/cm}^3 \times 6.7 \text{ cm}^3 = 75.71 \text{ g}$
3. $V = 18.3 \text{ g} \div 7.31 \text{ g/cm}^3 = 2.5 \text{ cm}^3$
4. $D = 5.4 \text{ g} \div 4.0 \text{ mL} = 1.35 \text{ g/mL} = 1.35 \text{ g/cm}^3$